Elements of Structures and Defects of Crystalline Materials

Elements of Structures and Defects of Crystalline Materials

Tsang-Tse Fang
National Cheng Kung University

ELSEVIER

Elsevier
Radarweg 29, PO Box 211, 1000 AE Amsterdam, Netherlands
The Boulevard, Langford Lane, Kidlington, Oxford OX5 1GB, United Kingdom
50 Hampshire Street, 5th Floor, Cambridge, MA 02139, United States

British Library Cataloguing-in-Publication Data
A catalogue record for this book is available from the British Library

Library of Congress Cataloging-in-Publication Data
A catalog record for this book is available from the Library of Congress

ISBN: 978-0-12-814268-4

Publisher: Matthew Deans
Acquisition Editor: Christina Gifford
Editorial Project Manager: Naomi Robertson
Production Project Manager: Vijayaraj Purushothaman
Cover Designer: Victoria Pearson

Typeset by MPS Limited, Chennai, India

Contents

Preface ix
Acknowledgements xiii

Part I
Structures of the Crystalline Materials

1. The Electron Configuration of Atoms

1.1 Atoms with a Single Electron 3
1.2 Atoms with More Than One Electron 6
1.2.1 Penetration and Shielding 7
1.2.2 Energetic d-Orbital Collapse of Free Neutral Atoms at the Beginning of the Transition Rows 8

2. Bonding Within Crystal Structures

2.1 Bonding in Ionic Crystals 13
2.1.1 Energy for Forming the Ions From Neutral Atoms 13
2.1.2 Thermodynamic Viewpoint of Bond Energy 15
2.2 Covalent Bonds 29
2.2.1 Formation Energy and Mechanism of Covalent Bonding 30
2.3 Metallic Bonds 31
2.3.1 Cohesive Energy of Metals 31
2.4 Effect of Bonding on the Material Properties by Means of the Potential-Well Concept 32
2.4.1 Elastic Modulus 32
2.4.2 Melting Temperatures 33
2.4.3 Thermal Property 33

3. The Structures of Crystalline Crystals

3.1 Arrangements of Atoms and Ions in Crystalline Solids: Space Lattice 37
3.2 Metallic Structures 39
3.3 Ionic Structures 40
3.3.1 Pauling's Rules 41

3.4 **Structural Distortion in Ionic Structures** 46
3.4.1 Effects of Crystal Field on the Structures of the Oxides 47
3.4.2 Crystal Field Stabilization Energy 51
3.4.3 Effects of the d Electron Configuration on the Radii of Transition Metal Ions 53
3.4.4 Jahn–Teller Distortions 55
3.4.5 Structure Distortions Arising from Asymmetric Electron Density or Inert Pair Effect 59
3.5 **Structure of Material Technological Interest** 59
3.5.1 Structure and the Related Feature of ZnO 61
3.5.2 Structure and the Related Feature of SiC 62
3.5.3 Structure and the Related Feature of TiO_2 63
3.5.4 Structure and the Related Feature of ZrO_2 63
3.5.5 Structure and the Related Feature of Spinel (AB_2O_4) 64
3.5.6 Structure and the Related Feature of Perovskite ABO_3 67

Part II
Defects of Crystalline Materials

4. Point Defects in Crystalline Materials

4.1 **Point Defects in Metals** 83
4.1.1 Solid Solutions 83
4.1.2 Factors Affecting the Solubility of Impurity Atoms in a Substitutional Solid Solution 85
4.1.3 Factors Affecting the Solubility of the Impurity Atoms in an Interstitial Solid Solution 85
4.1.4 Equilibrium Concentration of Point Defects 86
4.2 **Point Defects in Ionic Solids** 92
4.2.1 Notation for the Description of Point Defects in Binary Metal Oxides 92
4.2.2 Defect Structures in Stoichiometric Ionic Crystals 94
4.2.3 Defect Structures in Nonstoichiometric Ionic Crystals 94
4.2.4 Defect Reactions 95
4.2.5 Formulation of Defect Equations 96
4.2.6 Defect Equilibrium in Ionic Solids 109
4.2.7 Defect Association 126

5. Line Defects in Crystalline Solids

5.1 **The Discrepancy Between the Theoretical and Observed Yield Stresses of Crystals** 129
5.2 **Observations of Dislocations** 131
5.3 **Crystallographic Slip** 132

5.4 **Elementary Geometric Characteristics of Dislocations** 132
5.4.1 The Edge Dislocation 135
5.4.2 The Screw Dislocation 136
5.4.3 The Mixed Dislocation 136
5.5 **Critical Resolved Shear Stress** 137
5.6 **Plastic Flow Associated with the Slip of the Dislocation Motion** 138
5.7 **Stress Fields of Dislocations** 142
5.7.1 Constitutive Equations with Lame Constants 142
5.7.2 Stress Field of a Screw Dislocation 143
5.7.3 Stress Fields of an Edge Dislocation 145
5.8 **Energy of a Dislocation** 145
5.8.1 Strain Energy of a Screw Dislocation 146
5.8.2 Strain Energy of an Edge Dislocation 147
5.9 **Line Tension of a Dislocation** 147
5.10 **Forces Upon Dislocations** 148
5.11 **The Bowing of a Dislocation** 149
5.12 **Force Between Dislocations** 150
5.13 **Peach–Koehler Equation** 151
5.14 **Reactions Between Dislocations** 154
5.14.1 Dislocation Nodes 154
5.14.2 Kinks and Jogs 156
5.14.3 Intersections of Dislocations 157
5.15 **Extended Dislocations** 159
5.15.1 Partial Dislocations 160
5.15.2 Faults in the Stacking in FCC Crystals 161
5.15.3 Equilibrium Separation Between Partial Dislocations 162

6. Two-Dimensional (Interfaces) and Three-Dimensional (Second Phases) Imperfections in Solids

6.1 **Grain Boundaries** 169
6.1.1 Tilt Boundary 171
6.1.2 Twist Boundary 173
6.1.3 Coincidence Site Lattice 173
6.2 **Interphase Boundaries in Solids** 174
6.2.1 Coherent Boundary 174
6.2.2 Incoherent Boundary 175
6.2.3 Partially Coherent or Semicoherent Boundary 175
6.3 **Surface Tension, Surface Stress, and Surface Free Energy of Interfaces** 176
6.4 **Free Surface** 179
6.4.1 Calculate the Surface Free Energy of a Pure Solid FCC Metal 180

6.5 Interfaces of Phases 181
6.5.1 Wetting 181
6.5.2 Equilibrium Shapes of Grains 182
6.5.3 Morphological Changes of Second Phases 184
6.6 Effect of Interface Curvature on the Equilibrium Pressure Between Two Phases 188
6.7 Effect of the Interface Curvature on the Equilibrium Solubility Between Two Phases 191
6.8 Equilibrium Vacancy Concentration Changes at Curved Surfaces: Driving Force for Sintering 194

Appendix I: Spectroscopic Notation 199
Appendix II: Band Structures of Mott–Hubbard Materials and the Transition Metal Ionic Compounds, and Polaron 201
Bibliography 207
Index 211

Preface

As sparked by the discovery of the high T_c superconductor, growth and characterization of the materials mainly concerned with structures and defects are at the forefront of the research in materials science. The significance is further boosted on account of the subsequent discovery of the materials with so-called colossal properties, including magnetoresistance and dielectric constant. These materials are of perovskite or perovskite-like structures with transition metal ions on the B site and show an enormous variety of intriguing electronic or magnetic properties. This variety is not only related to their chemical flexibility concerning the defects, but also to a larger extent related to the complex character concerning the interplay between transition metal ions in certain coordinations and oxygen. Since these new emerging materials with multiple functions provide the most interesting and challenging topics in fundamental physics and the cutting edge of new technologies, they have drawn considerable worldwide attention of scientists with highly interdisciplinary backgrounds. However, research on these subjects is experiencing a superfast development and facing more completion. In response to these challenges, the scientists have to gain and grasp the desired knowledge in a short time. For most existing textbooks, they either are inadequate, focusing on one of the structures and defects, or are voluminous, including a wide variety of subjects. By contrast, this book is designed not only to be concise, yet comprehensive to thoroughly cover the fundamental principles of the structures and defects of crystalline materials, but also to provide deep insights into understanding the relationships of properties, defect chemistry, and processing of concerned materials. What is more, the issue concerning the significant effects on the properties of the functional oxides due to the crystal fields is missing from most of the existing related textbooks. This issue has been included in Section 3.4 of this book. In the fast-paced world, it is more important than ever to understand the fundamentals. Thus this book has been focused on the physics and the atomic mechanics of the problems to cover sufficient fundamentals, deep enough to allow the readers to delve into the most recent literature without too much difficulty. In addition, this book appeal to those other than materials science and engineering backgrounds, thus the author has made an effort to write this book so that it does not assume much prior knowledge of the subject from the readers. Furthermore, there are several issues in most existing textbooks that come from an "old" viewpoint. This book has updated these issues including extensions or corrections, e.g., Section 1.2.2 Energetic d-Orbital Collapse of Free Neutral Atoms at the Beginning of the Transition Rows, Section 2.3.1 Formation Energy and Mechanism of Covalent Bonding, Section 3.4.5 Structure Distortions Arising From Asymmetric Electron Density or Inert Pair Effect, Section 5.6 Plastic Flow Associated With the Slip of the Dislocation Motion, and Section 6.3 Surface Tension, Surface Stress, and Surface Free Energy of Interfaces. Finally, it has to be stressed that for most functional materials, to fully understand the interactions between properties and defect chemistry and processing, the knowledge concerning both structures and defects is necessary since they are correlated. This book has included the essence of the knowledge concerning the author's more than 30 years of research to provide deep insights into understanding the relationships of properties, defect chemistry, and processing.

This book is divided into two parts with six chapters. The first three chapters deal with the structures (Part I) and the remaining three chapters are concerned about the defects (Part II). Since the knowledge of the electron configuration of elements is necessary for understanding the nature of chemical bonding, it is discussed in the first chapter. This chapter is by no means intended to be comprehensive but to lend support to the aspects that are fundamental to the Section 3.4 of d electron concerned crystal fields. The determination of the material's structure is mainly from knowledge of its constituents' chemical bonding, which also has a great influence on the physical and chemical properties of materials. Chapter 2, Bonding Within Crystal Structures, thoroughly describes the bonding formation within the crystal structures of different materials, wherein the formation energy and mechanism of covalent bonding have been updated and the potential well concept has been established to correlate with the material properties. Chapter 3, The Structures of Crystalline Crystals, introduces the basic principles for constructing the material's structure. In view of the importance of the effects of the structure distortion on the material properties due to the crystal fields, the related topics have been included in Section 3.4. Moreover, several materials being still under intensive investigation have been illustrated to provide deep insights into understanding the effects of the relationships of processing, structures, and defects on the material properties.

The defects of materials are explored in Part II. Chapter 4, Point Defects in Crystalline Materials, deals with the point defects of metal and ceramics. While it starts from a very basic level, it proceeds to fairly advanced interpretations and covers the major fundamentals of defect structures and chemistry. While it is impossible to include all topics related to line imperfections, i.e., dislocations, Chapter 5, Line Defects in Crystalline Solids, essentially covers in ample detail the fundamentals of the characteristics of dislocations, wherein physics and the atomic mechanics of several issues have been described in detail. In view of the significant influence of the morphologies including size, shape, and distribution of grains and phases on the microstructure evolution and in turn the properties of materials, the final chapter focuses on the fundamentals of interface energies including single-phase (grain) boundary and interphase boundary, wettability determining the morphologies of phases, and the change of surface stress at a curved surface leading to the kinetic processes like Ostwald ripening and sintering. The distinction between surface energy, surface tension, and surface stress of liquids and solids in terms of atomic viewpoint is updated.

Acknowledgements

The accomplishment of this incredible task is by the grace of the Lord who lavishes wisdom, knowledge, and understanding upon me for the fear of the Lord is the beginning of wisdom and knowledge of the Holy One is understanding. Many thanks are to Prof. L.H. Chen for the helpful discussions and valuable suggestions concerning dislocations. It is also wished to acknowledge the generous assistance from Profs. J.C. Kuo and W.D. Hsu who provided me the computer-aided graphs of structures. Especially, I am most grateful to my wife for not only her support and patience during the preparation of this book but also her drawing most of the graphs in this book.

Part I

Structures of the Crystalline Materials

Concerning the structures of perfect crystalline solids, they possess a perfectly ordered, periodic arrangement of atoms or ions. The understanding of the structure of atoms would be a priori to describe the arrangement of atoms. This is necessary since the arrangement of atoms is dependent on the atom bonding whose origins are found to be in the electronic structure of atoms. The following chapter will deal with the electron configuration of atoms.

Chapter 1

The Electron Configuration of Atoms

1.1 ATOMS WITH A SINGLE ELECTRON

To appreciate the energetics and shapes of an atomic orbital is to solve the so-called Schrödinger's wave equation. The simplest case is the hydrogen atom whose electron behavior is described by the following usual form of the Schrödinger's wave equation:

$$\nabla^2\psi + \frac{8\pi^2 m_e}{h^2}(E - V)\psi = 0 \tag{1.1}$$

where ∇^2 is the Laplacian operator, expressed in Cartesian coordinates as

$$\nabla^2 = \frac{\partial^2}{\partial x^2} + \frac{\partial^2}{\partial y^2} + \frac{\partial^2}{\partial z^2}$$

E and V are the total energy and the potential energy of electron in a hydrogen atom, respectively; m_e is the electron mass. The wave function, ψ, is the solution of Eq. (1.1). Since the orbital of the hydrogen atom is spherically symmetrical, the description of the orbital would be easier to be tackled in spherical coordinates, as shown in Fig. 1.1. Eq. (1.1) can further be transformed in spherical coordinates as

$$\frac{1}{r^2}\frac{\partial}{\partial r}\left(r^2\frac{\partial\psi}{\partial r}\right) + \frac{1}{r^2 \sin\theta}\frac{\partial}{\partial\theta}\left(\sin\theta\frac{\partial\psi}{\partial\theta}\right) + \frac{1}{r^2\sin^2\theta}\frac{\partial^2\psi}{\partial\phi^2} + \frac{8\pi^2 m_e}{h^2}(E-V)\psi = 0 \tag{1.2}$$

The wave function now is a function of the three variables of r, θ, ϕ and can be represented by three separated functions $R(r)$, $\Theta(\theta)$, and $\Phi(\phi)$ as follows:

$$\psi(r,\theta,\phi) = R(r)\Theta(\theta)\Phi(\phi) \tag{1.3}$$

Substitute Eq. (1.3) into Eq. (1.2) and multiply throughout by $r^2\sin^2/R\Theta\Phi$, yielding

Elements of Structures and Defects of Crystalline Materials.
DOI: http://dx.doi.org/10.1016/B978-0-12-814268-4.00001-1

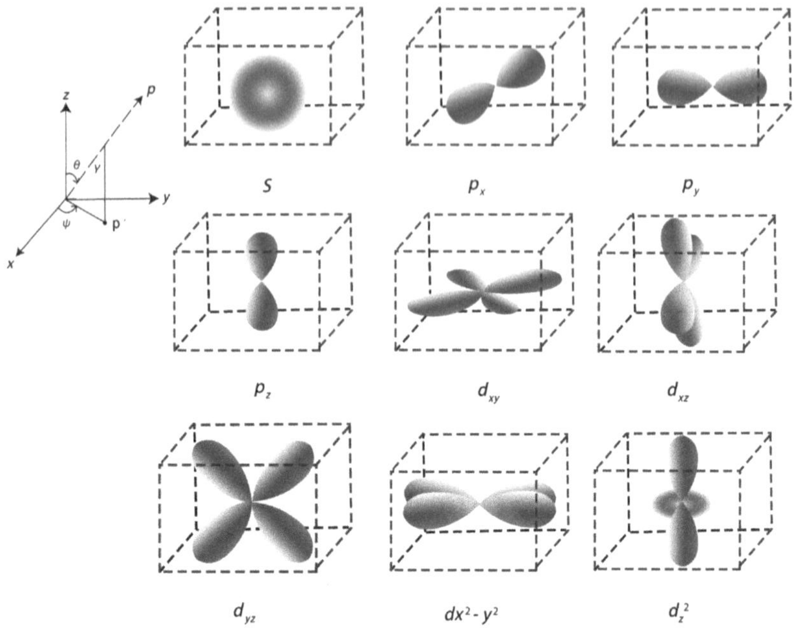

FIGURE 1.1 Setup of relationship between Cartesian coordinates, x, y, z, and the spherical polar coordinates r, θ, ϕ, and cubes based on the Cartesian axes are used to identify the axes of contour maps of the electron density distribution for s, p, and d orbitals.

$$\frac{\sin^2\theta}{R}\frac{d}{dr}\left(r^2\frac{dR}{dr}\right)+\frac{\sin\theta}{\Theta}\frac{d}{d\theta}\left(\sin\theta\frac{d\Theta}{d\theta}\right)+\frac{8\pi^2 m_e}{h^2}r^2\sin^2\theta(E-V)=-\frac{1}{\Phi}\frac{\partial^2\Phi}{\partial\phi^2} \tag{1.4}$$

Underlying Eq. (1.4), because both sides of this equation with different variables, the term of each side must be equal to the same constant m^2 for the equation to be valid.

For the term on the right, we obtain

$$\frac{1}{\Phi}\frac{\partial^2\Phi}{\partial\phi^2}+m^2=0 \tag{1.5}$$

For the terms on the left, the yielded equation is further divided throughout by $\sin^2\theta$ and rearranged to obtain an equation in R and Θ. Let each side of this equation be equal to the same constant α, giving the following two equations:

$$\frac{1}{R}\frac{d}{dr}\left(r^2\frac{dR}{dr}\right)+\frac{r^2 8\pi^2 m_e}{h^2}(E-V)-\alpha=0 \tag{1.6}$$

$$\frac{1}{\Theta\sin\theta}\frac{d}{d\theta}\left(\sin\theta\frac{d\Theta}{d\theta}\right)-\frac{m^2}{\sin^2\theta}+\alpha=0 \tag{1.7}$$

The acceptable (physically appropriate) solutions are further selected for Eq. (1.5), (1.6), and (1.7). The solution of Eq. (1.5) is

$$\Phi = \frac{1}{\sqrt{\pi}} \sin m\phi \tag{1.8}$$

For acceptable values of $\Phi(\phi)$, the values of the constant m are 0, ± 1, ± 2, . . ., etc., actually indicating the magnetic quantum number. For the possible solution of Eq. (1.7), we find out that $\alpha = l(l+1)$, where l can have the values of $|m|$, $|m|+1$, $|m|+2$, . . ., etc., and this l refers to the orbital quantum number. Another quantum number is further introduced, i.e., n, the so-called principal quantum number, for the solution of Eq. (1.6). The values of n can be taken as $l+1$, $l+2$, $l+3$, . . ., etc. The characteristics of these three quantum numbers are briefly reviewed as follows:

1. n: principal quantum number
   It specifies the major axis of an elliptical orbit and denotes the probable size of the "orbit" and the energy shell. The letters K, L, M, N, etc., corresponding to $n = 1, 2, 3, 4$, etc., respectively, are used in X-ray and spectroscopic work. The total possible number of electrons with a given value of n is $2n^2$.
2. l: orbital quantum number
   It is a measure of the angular momentum of the orbit motion of the electron and determines the shape of the orbital for any given value of n. It can only assume the values 0, 1, 2, . . ., $n-1$. The maximum number of states is given by $2(2l+1)$. All orbitals with $l = 0$ are called s orbitals (s: sharp), $l = 1$ are called p orbitals (p: principle), $l = 2$ are called d orbitals (d: diffused), and $l = 3$ are called f orbitals (f: fundamental).
3. m: magnetic quantum number
   In the presence of a magnetic field, m determines the spatial orientation of the plane of the "orbit." It varies in integral steps from $-$l to l, including zero.

To specify a particular energy state or quantum state for an electron, a fourth quantum number is required in addition to the aforementioned three quantum numbers. The fourth quantum number arises from the fact that the electron itself has an angular momentum referred to as the spin. The spin quantum number, s, ($s = \pm 1/2$), describes the angular momentum components of the spin and denotes the direction of electron spin.

The wave function of the hydrogen electron (Eq. (1.3)) can further be specified in terms of the quantum numbers as:

$$\psi_{n,l,m}(r,\theta,\phi) = R_{n,l}(r)\Theta_{l,m}(\theta)\Phi_m(\phi) \tag{1.9}$$

While $R^2_{n,l}(r)$ indicates the probability of finding the electron at a certain distance from the nucleus, the shape and orientation of the orbital are determined by $\Theta_{l,m}(\theta)$ and $\Phi_m(\phi)$. The conventional contour maps of the

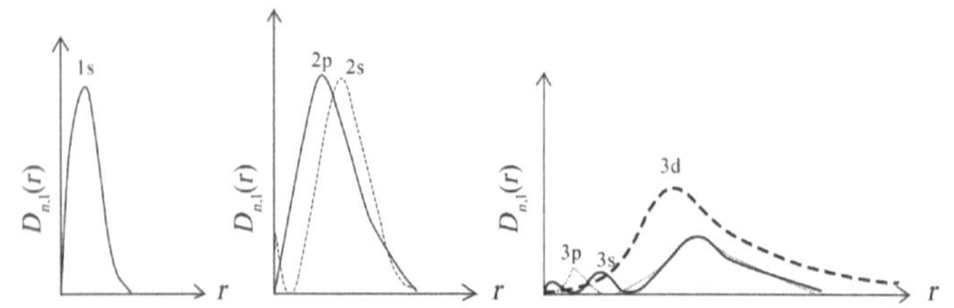

FIGURE 1.2 Radial distribution functions for orbitals of 1s, 2s, 3s, 3p, and 3d of the hydrogen atom, indicating the probability of finding the electron at a distance r from the nucleus for each atomic orbital and the comparison of the degree of penetration of the electron in the orbitals of different principal quantum numbers.

electron density distribution for s, p, and d orbitals are represented in Fig. 1.1. The s orbital independent of θ and ϕ reveals spherical symmetry but p and d orbitals depending on r, θ, and ϕ show nonspherical symmetry, indicating the directionality in their orbit. The probability to find an electron in a spherical shell is $4\pi r^2 R_{n,l}^2(r)dr$, where r is a distance from the nucleus, and the function $D_{n,l}(r) = 4\pi r^2 R_{n,l}^2(r)$ is known as the radial distribution or probability function. Fig. 1.2 shows the function $D_{n,l}(r)$ varying with r of orbitals 1s, 2s, 3s, 3p, and 3d.

For each orbital, the r value of each maximum probability corresponds to the radius of the corresponding Bohr orbit. Here, it should be pointed out that there is a finite probability at an infinite distance from the nucleus different than the classical concept of the Bohr orbit.

Concerning the electron structure of single electrons like hydrogen atoms, the energy states can basically be solved from Eq. (1.6) by substituting V with $-Ze^2/r$ (where Z is atomic number; e is electron charge; and r is the distance of the electron from the nucleus) and expressed as:

$$E_n = -\frac{2\pi Z^2 e^4 m_e}{n^2 h^2} \qquad (1.10)$$

Thus for the single electron case, the energy of each subshell is determined uniquely by the principal quantum number n, i.e., the 2s and 2p subshells are of identical energy and the 3s, 3p, and 3d are of identical energy. Its values are typically between 1 eV and 10^3 eV. However, for multi-electron atoms, electron–electron interactions should be taken into account, which will be discussed in the following section.

1.2 ATOMS WITH MORE THAN ONE ELECTRON

If there is more than one electron around the atom, interactions between electrons would raise the energy states. Three kinds of energy contributions to polyelectronic atoms have to be taken into account: (1) the kinetic energy of electrons moving around the nucleus; (2) the attractive energy between electrons and nucleus; and (3) the repulsive energy between electrons.

However, because of the uncertainty of the pathways of the electrons, the repulsive energy between electrons cannot be exactly evaluated, which is known as electron correlation problems. Nevertheless, this has been approximated to treat each electron as if it were moving in a field established by the net electrostatic force between the attraction of the nucleus and the average repulsions of the rest electrons.

While for the single electron case, the energy of subshells determined uniquely by the principal quantum number n is degenerate, that is, the energy of 2s and 2p subshells and the 3s, 3p, and 3d are identical, respectively, in multi-electron atoms, orbitals with the same n but different ℓ have different energies. The reason for this is that the lower angular momentum s orbital is characterized by an electron distribution closer to the nucleus, and thus the electron–nuclear attraction becomes more important, while for higher angular momentum, it results in greater radial distances from the nucleus for a given principal quantum number n, leading to a smaller and smaller effective nuclear charge as we progress from p to d to f, namely, electron–electron repulsion becomes more significant. Since the competitive balance between electron–nuclear attraction and electron–electron repulsion determines the energy of the individual orbitals in the structure of any atom, understanding of this balance is necessitated, which essentially can be realized in terms of two viewpoints of penetration and shielding.

1.2.1 Penetration and Shielding

Penetration is considered to be a measure of how close an electron comes to the nucleus, which would become significant since the potential energy becomes more negative with increasing Z and decreasing r. Shielding, however, is a measure of how electrons that occupy a spatial domain closer to the nucleus serve to reduce the attraction between the nucleus and electrons at greater distances from the nucleus. While the interactions between electrons are often ignored if the spatial overlapping of the electron wave functions is low, in most cases, a shielding effect would arise from interactions between electrons, where the outer electrons see an effective nucleus of reduced charge, since the inner electrons are bound tightly to the nucleus and partially cancel its charge. This leads to a correction of Z in Eq. (1.10) by substituting with an effective nuclear charge, Z_{eff}, that depends strongly on the principal quantum number:

$$E_n = -\frac{2\pi {Z_{\text{eff}}}^2 e^4 m_e}{n^2 h^2} \tag{1.11}$$

The effective nuclear charge, Z_{eff}, can be defined as the number of protons in the nucleus, Z, minus the average number of electrons, S, between the nucleus and the electron of interest.

It should be stressed here that both effects, penetration and shielding, are deeply influenced by the radial density of the probability of finding an electron at a given radius. For instance, as observed in Fig. 1.2, the presence of the radial node in the 2s orbital means that there is a region of space relatively close to the nucleus in which the 2s electron is likely to be found. In effect, an electron in the 2s orbital will spend more time nearer to the nucleus than an electron in a 2p orbital. As a result, the 2s orbital is of greater negative energy (greater stability, lower energy) than the 2p orbital. In addition, in comparing quantitatively the binding energy of a single electron in He^+ and an electron in the filled $1s^2$ orbital of He, the energy required to remove one electron decreases from 5250 kJ/mol for the former to 2372 kJ/mol for the latter as a consequence of both shielding and penetration. Furthermore, upon filling the electrons into the orbitals, there are two advantages for the energy state of 2s to be lower than that of 2p: reducing the potential energy by decreasing r via the penetration and reducing the shielding of 1s electrons.

Concerning the energy order for orbitals with $n = 3$, in analogy to the arguments relevant to 2s and 2p, it would also be in the order of $3s < 3p < 3d$, in view of the raidial probability function varied with the distance r from the nucleus, observed in Fig. 1.2. However, in comparing 3d to 4s, there is energetic d-orbital collapse, that will be discussed in the following section.

1.2.2 Energetic d-Orbital Collapse of Free Neutral Atoms at the Beginning of the Transition Rows

In view of the recent significant findings like high-T_c superconductors, colossal properties of magnetoresistance ($(LaSr)MnO_3$) and dielectric constant ($CaCu_3Ti_4O_{12}$), and multiferroics ($BiFeO_3$), transition elements acting either as matrix or impurity ions have a tremendous influence on these remarkable properties. Thus, a full understanding of the electron structures of these elements is necessary.

For the 3d orbital, one would expect the 3d orbital to be lower in energy (a more negative energy) than the 4s orbital in terms of the principal quantum number n. However, as observed in Fig. 1.3, by comparison of the radial distribution functions of 3d and 4s orbitals, it is apparent that the 4s orbital

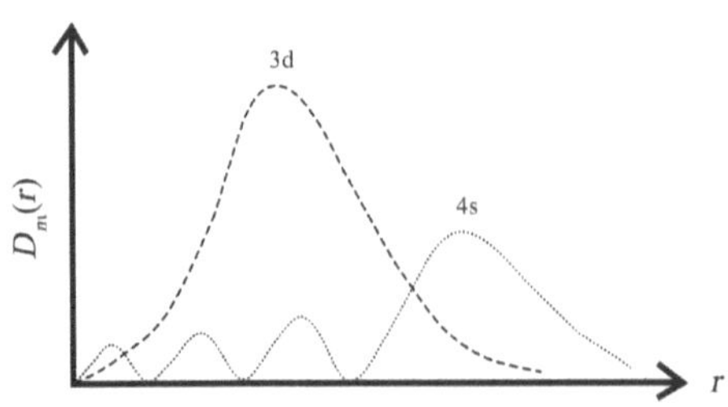

FIGURE 1.3 The comparison of the radius distribution function 3d, and 4s atomic orbital, indicating the diffuseness and higher degree of penetration of the 4s orbital.

is more penetrating than the 3d orbital, leading to the decrease in potential energy and in turn making the energy of the 4s orbital lower.

However, it is known from vacuum atom-spectroscopic data that the order of the orbitals of Ca is indeed $3p \ll 4s < 3d < 4p \ll$, etc., whereas that for Sc and subsequent transition elements is $3p \ll 3d < 4s < 4p \ll$, etc., which is the so-called "d-orbital collapse." In general, it is the result of the interplay of nuclear attraction, of imperfect shielding by the inner-core electrons, and of angular momentum-dependent centrifugal forces. Nevertheless, d-orbital collapse will further be clarified in the following. Going from Ca to Sc indicates that you are adding one more proton to the nucleus. This makes the nuclear charge larger and therefore both the 4s and 3d orbitals are stabilized (their energies decrease). However, the energy of the 4s orbital decreases more slowly than that of the 3d orbital, because the 4s orbital is relatively radially diffuse, and its maximum in the radial distribution function occurs at a larger value of *r*. Additionally, the poor shielding of the 3d orbitals would make the 3d orbital energy decrease faster. Therefore, it is expected that the orbital energy of 3d would be lower than that of 4s. However, upon filling the electrons, the order to fill the subshells of 3d and 4s is not what we expect in accordance with the orbital energy. For Sc, the electrons preferentially occupy the 4s subshell even though it is higher in energy. This is because the 3d shell is rather compact, the electron repulsion in the d shell increases strongly with increasing d occupation in contrast to the diffuse Rydberg 4s orbital, in which the electrons have more space and experience less repulsion. A remarkable example is that as one electron is added to a V^+ ion with the electron configuration of $3d^44s^0$, it becomes a V^0 atom whose electron configuration is $3d^34s^2$ instead of $3d^44s^1$. Thus an empty 4s orbital in Sc has a higher energy than an empty 3d orbital, but a filled 4s orbital has a lower energy than a filled 3d orbital. The fact that 4s >3d in energy also explains why, for the transition metals, the 4s electrons are removed first upon ionization.

In constructing the electron configuration of polyelectronic atoms, the so-called aufbau principle is adopted, i.e., electrons are added one by one to the orbitals in a specified energy order. Upon filling the electrons into the orbitals, in addition to the knowledge concerning the energy ordering of orbitals deduced from the aforementioned penetration, shielding, and electron–electron repulsion, the conceptual basis including two primary considerations is required: (1) The Pauli exclusion principle requires that a given energy or quantum state specified by four quantum numbers, *n*, *l*, *m*, and *s* be occupied by one electron; alternatively, in an energy state, no two electrons can exist having all four quantum numbers, *n*, *l*, *m*, and *s*, the same. This principle also suggests that as the number of electrons increases, they are forced to occupy a higher energy level, namely, higher *n* values. (2) In a partially filled orbital, the number of electrons having the same spin is necessary to be maximum, known as Hund's first rule or the maximum multiplicity principle. This rule recognizes that for degenerate orbitals with the same *n* and *l* quantum numbers but with different *m* values, the electron–electron repulsion requires that orbitals with different *m* are filled with unpaired electrons before the electrons pair up in the same orbital.

It is noticed that the electron configurations of Cr and Cu are always considered to be exceptional in the 3d transition metals, where the former is usually considered to be with a half-filled subshell ($3d^5$) and the latter a fully-filled 3d subshell ($3d^{10}$). While the related reasons are too complicated to be substantially explicated, they are usually ascribed to the symmetry and exchange energy, which however is misleading (Wang et al., 2006; Schwarz, 2010). The ground states of free atoms are determined by the Coulomb and spin–orbit couplings of the open *n*, *l* orbital shells. Nevertheless, a chemically rather unimportant small spin–orbit energy shift by a few kJ mol^{-1} can result in a different atomic ground level and a different corresponding configuration because of the close energy separation of the 4s and 3d subshells, for instance, Cr or Ni. For Cr, it is expected that the lowest average configuration energy corresponds to $3d^44s^2$ on account of the high repulsion of electrons in 3d orbitals, which is indeed evidenced by the spectroscopic data showing that as one electron is added to a Cr^{1+} ion with the electron configuration of $3d^54s^0$, it becomes a Cr^0 atom whose electron configuration is $3d^44s^2$ rather than $3d^54s^1$. The $3d^54s^1$ average is $\sim$0.66 eV above the $3d^44s^2$ ground configuration whose J level $^5D_0$ (spectroscopic notation is shown in Appendix I) is stabilized by $\sim$2.5 eV. However, the lowest J level $3d^54s^1$ ($^7S_3$) is split down by more than 4 eV, and thus forms the ground level of Cr. It is implied that essentially, half-filled shells are not particularly stable, but they give rise to a particularly low atomic LS coupling. Moreover, since the energies of the two configurations $3d^44s^2$ and $3d^54s^1$ of Cr are close, the two configurations are strongly mixed for several states. As for Ni, the ground configuration of neutral Ni in chemical substances is $3d^{10}$; whereas the lowest configuration of free Ni^0 atoms is $3d^94s^1$ rather than $3d^84s^2$ as expected, which is about 100 kJ/mol higher! However, since the small spin–orbit splitting of $3d^84s^2$ (^3F) with two d holes ($\pm$13 kJ/mol) is slightly larger than that of $3d^94s^1$ (^3D) ($\pm$10 kJ/mol), the lowest J level $^3F_4$ of $3d^84s^2$ is eventually below the $^3D_3$ of $3d^94s^1$. Therefore, the basic valence configuration of Ni is defined as $3d^84s^2$ rather than the lower $3d^94s^1$ configuration in most chemistry textbooks. It is concluded that in the case of two different configurations with their LS couplings being energetically close, a small spin–orbit coupling may invert the order of the lowest configurations.

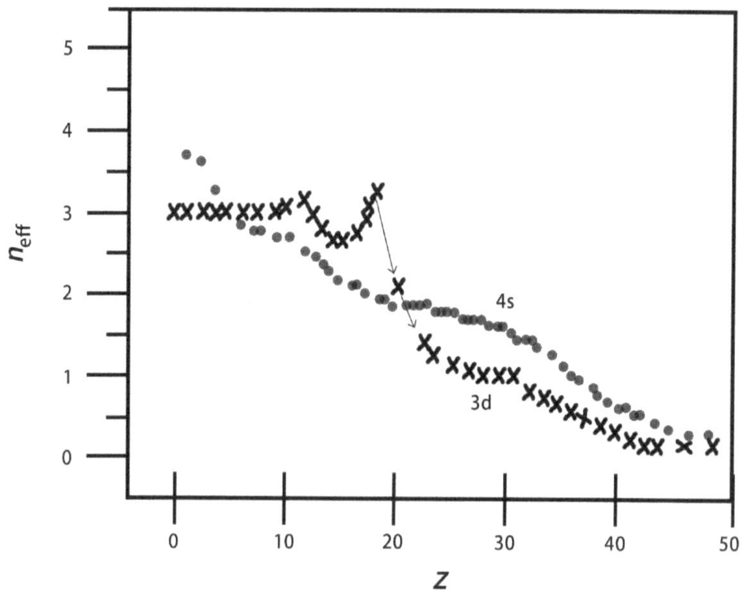

FIGURE 1.4 Energetic d-orbital collapse (indicated by arrows) of free neutral atoms at the beginning of the transition rows. Z is the nuclear charge. The orbital energies are represented by the effective quantum numbers $n_{eff} = \frac{1}{\sqrt{|\varepsilon|/13.6}}$. *From Schwarz, W.H.E., "The Full Story of the Electron Configurations of the Transition Elements" 2010. J. Chem. Educ. 87 (4), 444.*

Energetic d-orbital collapse of free neutral atoms at the beginning of the transition rows is shown in Fig. 1.4.

The established electron configuration of the atoms would further lend support to understand the interactions of atoms, namely, bonding, described in the next chapter.

Chapter 2

Bonding Within Crystal Structures

Bond energy is a result of atoms combining to form stable molecules and condensed phases that leads to a net lowering of the energy in the bonded atoms, compared to the set of isolated atoms. Crystal bonding determined by the characteristics of the outermost valence electrons has a significant influence on the physical and chemical properties of materials. Thus, in order to correlate structure and physical properties with the chemical composition of crystalline materials, it is necessary to have an adequate description of the outermost valence electrons. The characteristics of the outermost valence electrons depend on the influence of the nuclear charge, which varies according to which orbitals they occupy and to what degree they shield by the electrons closer to the nucleus. It is imperative to explore the nature of the forces that hold atoms together. The strong bonding gives rise to *primary bonds*, which can be metallic, ionic, and covalent, whereas the weaker bonding, including van der Waals and hydrogen bonds, is referred to as *secondary bonds*. Concerning crystalline solids, hydrogen bonds will be excluded from the discussion.

2.1 BONDING IN IONIC CRYSTALS

While there is no completely ionized bond, Pauling set up a scale of relative electronegativity (EN) defined as the power of an atom to attract electrons to itself to give a very rough guide for distinguishing predominantly ionic ($\Delta X > 1.7$) from predominantly covalent ($\Delta X < 1.7$) bonds, where ΔX is the EN difference. For LiF, EN of Li is 0.98 and F is 3.98. Since $\Delta X = 3$ is larger than 1.7, the bond of LiF can be considered as ionic. And the electrons are tightly bound to their respective atoms inside the crystal.

2.1.1 Energy for Forming the Ions From Neutral Atoms

Ionic bonds develop between strongly electropositive elements and strongly EN elements. Formation of the ions from neutral atoms is illustrated in Fig. 2.1, which can be evaluated by means of ionization energy (IE) and electron affinity (EA). The former is defined as the energy

Elements of Structures and Defects of Crystalline Materials.
DOI: http://dx.doi.org/10.1016/B978-0-12-814268-4.00002-3

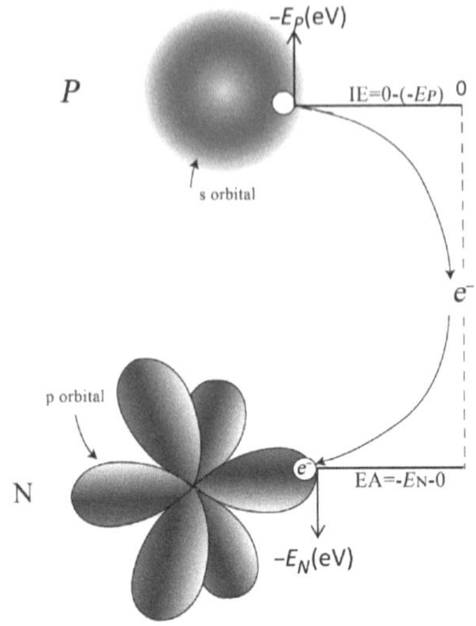

FIGURE 2.1 Overall energy changes in the formation of a pair of ions from the free atoms, and the formation of an ionic bond with respect to the ionization and affinitive energy changes.

required to completely remove a valence electron from the orbital of an isolated atom to an infinite distance; the latter is defined as the energy given off when an electron is added to the valence shell of an atom. The energy variation of interactions between the valence electrons of two atoms is represented in Fig. 2.1, in which the energy at the infinite distance is taken to be zero.

For the electropositive atom (P), assume the orbital energy of the valence electron is $-E_P$, being ionized, and thus $\mathrm{IE} = 0 - (-E_P) = E_P$. As this released electron is drawn into the EN atom, it will lower its energy from zero to the orbital energy of $-E_N$. Thus,

$$\mathrm{EA} = -E_N - 0 = -E_N$$

The net change in the energy of this electron is

$$\Delta E = -E_N - (-E_P) = E_P - E_N$$

Usually, IE > EA, so that $\Delta E > 0$, indicating the energy required to form the ions from neutral atoms.

According to the EN difference, LiF is considered to be ionic bonding, which can be set as an example to evaluate ΔE. Taking into account the following electron configurations of Li and F:

$$\mathrm{Li{:}1s^2 2s^1, F{:}1s^2 2s^2 2p^5}$$

The pertinent reactions of the interactions of electrons between Li and F are:

$$\mathrm{Li} + 5.4\ \mathrm{eV} \rightarrow \mathrm{Li}^{+} + \mathrm{e}^{-}(\mathrm{IE})$$

$$\mathrm{F} + \mathrm{e}^{-} \rightarrow \mathrm{F}^{-} + 3.6\ \mathrm{eV}(\mathrm{EA})$$

$$\mathrm{Li} + \mathrm{F} + 1.8\ \mathrm{eV} \rightarrow \mathrm{Li}^{+} + \mathrm{F}^{-}$$

Thus in this case $\Delta E = 1.8$ eV. The resulting ions are capable of forming ionic bonds with the energy (a net decrease in energy) arising from the electrostatic attraction of the ions.

2.1.2 Thermodynamic Viewpoint of Bond Energy

Consider the Gibbs free energy G, defined in terms of the internal energy E and entropy S of the system: $G = E - TS + PV$, where T and P are absolute temperature and pressure, respectively. For the solids, PV can be neglected if P is not high enough. Thus G can be replaced by F (Helmholtz's free energy, known also as cohesive energy or binding energy), giving $F = E - TS$. Furthermore, the internal energy of a crystal can be considered to be consisted of the lattice energy (U_L), with respect to the potential energy of the immobile ions in the equilibrium positions, and the vibration energy (E_V), concerning the energy of the mobile ions oscillating around the equilibrium position, i.e., $E = U_L + E_V$. If the temperature is not high, TS can be neglected and $E_V << U_L$, so $F \sim U_L$. In other words, the bond energy of a crystal can be predominantly determined by the lattice energy, which mainly involves the following energies:

1. Electrostatic energy
2. Repulsive energy
3. The repulsion energy between fully occupied orbitals at any attempt to superimpose them (Pauli principle)
4. Energy due to crystal field stabilization
5. Energy due to covalent bonds
6. Energy due to van der Waals force
7. Zero-point energy

We will discuss the main contributions including (1), (2), (6), and (7) to the bond energy in the following. Energy due to crystal field stabilization will be detailed in Section 3.4.1 of chapter 3.

2.1.2.1 Electrostatic Energy

Ionic bonds arise from the electrostatic interactions of the ions. The electrostatic potential energy between ions can be assumed to be replaced by appropriate point charges. The Coulomb interaction is described as follows. For international system (SI) units,

$$F = \frac{Z \cdot Z' \cdot e^2}{(4\pi\varepsilon_0)r^2} \tag{2.1}$$

where Z and Z' are charges. For centimeter−gram−second (CGS) units,

$$F = \frac{Z \cdot Z' \cdot e^2}{r^2} \tag{2.2}$$

In this book, CGS units will be adopted. Thus electrostatic potential energy can be expressed as

$$E_p = -\int_{\infty}^{r} F \cdot dr = \frac{Z \cdot Z' \cdot e^2}{r} \tag{2.3}$$

The computation of the energy of a cluster of ions approximated as point charges requires several steps:

Step 1: Calculate the total interaction energy, with the use of an equation similar to Eq. (2.3) between a reference ion and the other ions in the cluster; the interaction is given a plus or minus sign depending on whether the ions have the same or opposite charges.
Step 2: Repeat this summation for all the other ions in the cluster.
Step 3: Divide the calculated total energy by two, as each ion will be counted twice.

The energy so derived is called the Madelung energy of the cluster, which will be extended to the crystal structures where the ions will be approximated as the point charges. A structure with ions having a single valence electron, such as sodium chloride (NaCl), would serve as a good example. The structure of NaCl is a rocksalt structure shown in Fig. 2.2.

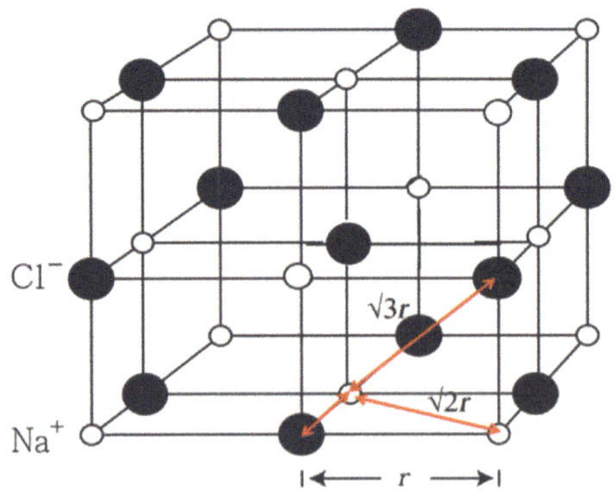

FIGURE 2.2 Structure of NaCl, where the interatomic distances of the neighboring atoms are specified.

Consider one particular ion, e.g., Na^+, and evaluate the interactions between it and its neighbors. It has six nearest neighbors Cl^- at a distance r. The attractive potential energy is

$$E_{p1} = -6 \times \frac{e^2}{r} \tag{2.4}$$

The next nearest neighbors to Na^+ are $12Na^+$ at a distance $\sqrt{2}r$, and this gives a repulsive potential energy:

$$E_{p2} = 12 \times \frac{e^2}{\sqrt{2}r} \tag{2.5}$$

The third nearest neighbors are $8Cl^-$ at a distance $\sqrt{3}r$, which is an attractive potential energy:

$$E_{p3} = -8 \times \frac{e^2}{\sqrt{3}r} \tag{2.6}$$

Continuing the calculation of the interaction energy for the further neighbors in the lattice and following the aforementioned steps concerning the calculation of the Madelung energy of a cluster of point charges, the Madelung (or electrostatic) energy per molecule of NaCl crystal is given by

$$\begin{aligned} U_a &= -\frac{e^2}{r}\left(6 - \frac{12}{\sqrt{2}} + \frac{8}{\sqrt{3}} - \frac{6}{\sqrt{4}} + \cdots\right) \\ &= -\frac{e^2}{r} \cdot A \end{aligned} \tag{2.7}$$

where A is the Madelung constant depending on geometry of the structure. In a structure in which the ions have valences of more than one, such as a compound M_aX_b, Eq. (2.7) would be revised as

$$U_a = -\frac{e^2}{r} \cdot A' \cdot Z_M \cdot Z_X \frac{a+b}{2} \tag{2.8}$$

where Z_M and Z_X are valences of the cation and anion, respectively; r is the shortest interionic distance between cation and anion; and A' is the reduced Madelung constant. Notice that Eq. (2.8) has taken the charge neutrality of $aZ_M = bZ_X$ into consideration. Nevertheless, conventionally, Eq. (2.8) is simplified into the form of Eq. (2.7), making

$$A' \cdot Z_M \cdot Z_X \frac{a+b}{2} = A \tag{2.9}$$

and thus A is called the conventional Madelung constant. The values of Madelung constants for different crystal structures are listed in Table 2.1.

TABLE 2.1 Conventional and Reduced Madelung Constants of Several Crystal Structures

Crystal Structure	Reduced Madelung Constant, A'	Conventional Madelung Constant $A = [Z_M Z_X(a+b)A']/2$
NaCl	1.7476	1.7476
CsCl	1.7627	1.7627
Zinc blende	1.6381	6.5524
Wurtzite	1.6413	6.5652
Fluorite	1.6796	5.0388
Rutile	1.6053	19.264
Corundum	1.6688	25.032

The following relationship is also used in some cases:

$$U_a = -\frac{e^2}{r} \cdot A'' \cdot Z^2 \tag{2.10}$$

where Z is the highest common factor of valence charges of the anion and cation, e.g., 1 for NaCl, CaF_2, and Al_2O_3, 2 for MgO, TiO_2, and ReO_3, etc.

2.1.2.2 *Repulsive Energy*

As the ions approach each other, the like charges repel and prevent the ions from coming any closer. Born proposed the form of repulsive energy as follows:

$$U_r = \frac{B}{r^n} \tag{2.11}$$

where B is an empirical constant and the Born exponent, n is usually in the range of 5–12.

2.1.2.3 *Approximation of Lattice Energy and Forces Between Ions in an Ionic Crystal*

The lattice energy of the crystal per molecule can be roughly evaluated by summing Eqs. (2.7) and (2.11):

$$U_L = -\frac{e^2 A}{r} + \frac{B}{r^n} \tag{2.12}$$

Eq. (2.12) can be graphically illustrated in Fig. 2.3A, where ΔE is defined in Section 2.1.1. The minimum in the curve corresponds to the equilibrium situation. At this minimum, it indicates that the interatomic distance is at equilibrium.

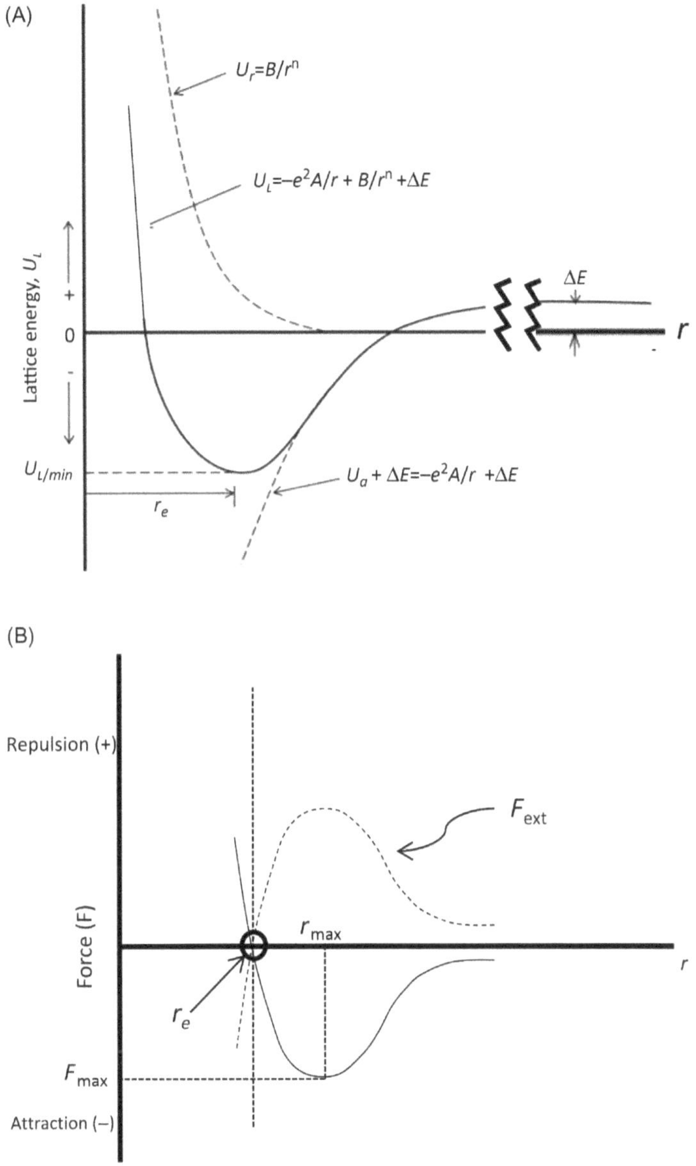

FIGURE 2.3 (A) Variations of attractive, repulsive, and lattice energies with interatomic distance for an ionic bond, where r_e is the equilibrium distance. (B) Corresponding force versus interatomic distance curve derived from curve (A), and the inverted diagram (dashed line) representing the external force versus interatomic distance curve.

The depth of the energy well of U_L at the equilibrium interatomic distance can be obtained by differentiating Eq. (2.12) with respect to r in the following:

$$\frac{dU_L}{dr} = \frac{e^2A}{r^2} - \frac{nB}{r^{n+1}} \tag{2.13}$$

Since underlying Fig. 2.3A,

$$\frac{dU_L}{dr} = 0 \quad \text{at } r = r_e \text{ (where } r_e \text{ is the equilibrium distance),}$$

we obtain

$$B = \frac{e^2 A r_e^{n-1}}{n} \tag{2.14}$$

Substitute Eq. (2.14) into Eq. (2.12) at the equilibrium distance to obtain the depth of the energy well corresponding to the bond energy:

$$U_{L,\min} = -\frac{e^2A}{r_e}\left(1 - \frac{1}{n}\right) \tag{2.15}$$

Due to the fact that the values of r_e and n can be obtained from X-ray crystallography and compressibility measurement, respectively, the lattice energy at the equilibrium distance can be evaluated.

Notice that if one mole molecules of the crystal are taken into account, the lattice energy may be written as

$$U_L = -\frac{e^2NA}{r} + \frac{NB}{r^n} \tag{2.16}$$

where N is Avogadro's number.

Since the material properties are closely related to the forces between the atoms, it is important to understand the variation of the forces with the interatomic distance. The net force between the ions can be obtained by differentiating Eq. 2.12:

$$F = -\frac{dU_L}{dr} = -\frac{e^2 \cdot A}{r^2} + \frac{n \cdot B}{r^{n+1}} \tag{2.17}$$

The corresponding force versus interatomic spacing graph is shown in Fig. 2.3B, where at $r > r_{\max}$, the attractive force dominates (negative) and the minimum at $r_{\max}$ corresponds to the maximum attractive force; at $r_e < r < r_{\max}$, the repulsive force starts to be involved; at $r < r_e$, the repulsive force prevails (positive). It is noticed that it can be nicely approximated as linear in the vicinity of r_e, where Hooke's law can be applied. On applying the external tensile force, the atoms are pulled apart causing the development of the attractive restoring force between them tending to pull them back together. The maximum attractive force can be obtained as follows.

$$\frac{dF}{dr} = \frac{2e^2 \cdot A}{r_{max}^3} - \frac{(n+1)n \cdot B}{r_{max}^{n+2}} = 0 \tag{2.18}$$

where

$$B = \frac{2e^2 \cdot A \cdot r_{max}^{n-1}}{(n+1)n} \tag{2.19}$$

Substitute Eq. (2.19) into (2.17) to obtain the maximum attractive force

$$F_{max} = -\frac{e^2 \cdot A}{r_{max}^2}\left(1 - \frac{2}{n+1}\right) \tag{2.20}$$

The theoretical strength can be considered to be a force as a function of the atom position shifting away from the equilibrium position, thus F_{max} can be considered to be the force that the maximum external force has to overcome in order to pull the crystal apart, i.e., break the bonds. As for the external force (or stress), it is defined as

$$F_{ext} = +\frac{dU_L}{dr}$$

Underlying the sign convention, the external tension force tending to increase the interatomic distance is positive but the external compressive force reducing the interatomic spacing is negative. Thus the maximum external force (F_{mext}) is $-F_{max}$. It is further assumed that the maximum position r_{max} can be considered as the distance at which most bonds will break as the material is sheared. Thus $r_{max} \approx \alpha r_e$, where α is a numerical value. It is also assumed that $(F_{mext}/r_e^2) \approx \sigma_{max}$, therefore, it is obtained

$$\sigma_{max} \approx \frac{e^2 \cdot A}{\alpha^2 r_e^4}\left(1 - \frac{2}{n+1}\right) \tag{2.21}$$

This can be regarded as the theoretical strength of the solid. Furthermore, as previously mentioned, the response of the region in the vicinity of r_e obeys Hooke's law. Thus it is reasonable to make the following approximation:

$$F_{ext} = k(r - r_e)$$

where k is the force constant or the stiffness of the bond, which is defined as

$$k = \left(\frac{dF_{ext}}{dr}\right)_{r=r_e}$$

In order to relate k with modulus, it is further assumed that stress (σ) can be represented as F_{ext}/r_e^2 and the strain (ε) as $(r - r_e)/r_e$ and by comparing with the relationship $\sigma = M\varepsilon$, where M is the modulus, yielding

$$M = \frac{k}{r_e}$$

Therefore,

$$M = \frac{1}{r_e}\left(\frac{dF_{\text{ext}}}{dr}\right)_{r=r_e}$$

$$M = -\frac{2e^2 \cdot A}{r_e^4} + \frac{(n+1)n \cdot B}{r_e^{n+3}} \tag{2.22}$$

By using the relationship of Eq. (2.19) and $r_{\max} \approx \alpha r_e$, we obtain

$$M = \frac{2e^2 \cdot A}{r_e^4}(\alpha^{n-1} - 1) \tag{2.23}$$

Finally, combine Eqs. (2.21) and (2.23) to eliminate A, giving

$$\sigma_{\max} \approx M \frac{1 - (2/(n+1))}{2\alpha^2(\alpha^{n-1} - 1)} \tag{2.24}$$

In general, the values of n are between 5 and 12, and the value of α is around 1.22 for ionic solids. As per Eq. (2.24), $\sigma_{\max} \approx (M/5)$ to $(M/28)$. Taking the average value of $n \approx 9$, $\sigma_{\max} \approx (M/15)$. Thus it may be concluded that the theoretical strength of a solid should be roughly one-tenth of its elastic modulus. (See also in Sec. 5.1)

2.1.2.4 van der Waals Attractive Energy

The van der Waals forces arise from nonsymmetrical charge distributions. The most important component of these forces can be ascribed to the interactions of electric dipoles. An electrical dipole is formed by a pair of oppositely charged particles separated by a small distance.

2.1.2.4.1 Field Produced by a Dipole

The electric potential of a point charge is expressed below, whose values are positive or negative depending on the sign of q.

$$V = \frac{q}{(4\pi\varepsilon_0)r} \tag{2.25}$$

If there is a charge q' at the distance r, the potential energy of charge q is

$$E_p = \frac{qq'}{(4\pi\varepsilon_0)r} \tag{2.26}$$

In the case of an electrical dipole, let the separation distance be a, and d_1 and d_2 be the respective distances from the two charges of the dipole to a point P with a positive test charge in space at a distance r from the midpoint of the dipole, as shown in Fig. 2.4. The resultant electric field exerted on the test charge by the dipole is also represented in Fig. 2.4.

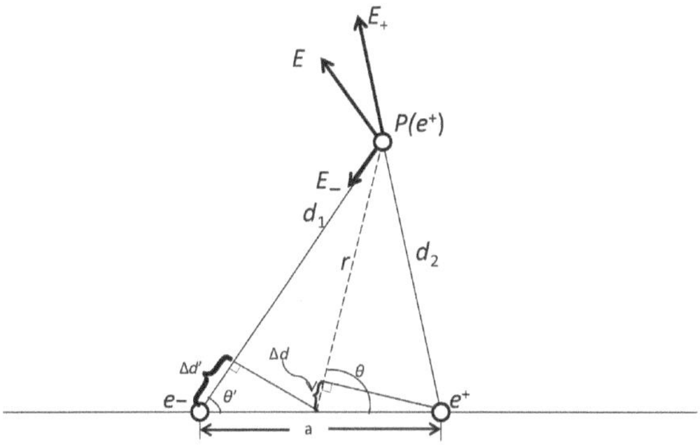

FIGURE 2.4 Electrostatic field produced by an electrical dipole.

There are several approximations, which can be made due to the extremely small distance between positive and negative charges in a dipole. Draw two lines from e^+ and midpoint vertical to line r and line d_1, respectively. According to Fig. 2.4,

$$r = d_2 \sin\theta + \Delta d = d_2 \sin\theta + \frac{a}{2}\cos\theta \tag{2.27}$$

For a trigon with an angle equal to 90°, the following relationship should be valid:

$$d_2^2 = (d_2 \sin\theta)^2 + \left(\frac{a}{2}\sin\theta\right)^2$$

Since $((a/2)\sin\theta)^2$ is very small,

$$d_2 \approx d_2 \sin\theta \tag{2.28}$$

In comparing Eqs. (2.17) and (2.28), we obtain

$$d_2 \approx r - \frac{a}{2}\cos\theta \tag{2.29}$$

In the same way,

$$d_1 = r\sin\theta' + \Delta d' = r\sin\theta' + \frac{a}{2}\cos\theta',$$

which can be approximated as

$$d_1 \approx r + \frac{a}{2}\cos\theta'$$

On account of the minute distance a, $\theta \approx \theta'$, and

$$d_1 \approx r + \frac{a}{2}\cos\theta. \tag{2.30}$$

Underlying the above approximations of Eqs. (2.29) and (2.30), the electrical potential at P is given by

$$V = \frac{e}{d_2} - \frac{e}{d_1} = \frac{e}{r - \frac{a}{2}\cos\theta} - \frac{e}{r + \frac{a}{2}\cos\theta} \tag{2.31}$$

$$V = \frac{e}{r}\left[\frac{1 + \left(\frac{a}{2r}\right)\cos\theta - 1 + \left(\frac{a}{2r}\right)\cos\theta}{1 - \left(\frac{a}{2r}\cos\theta\right)^2}\right] \tag{2.32}$$

$$V = \frac{e}{r}\left[\frac{\frac{a}{r}\cos\theta}{1 - \left(\frac{a}{2r}\cos\theta\right)^2}\right] \tag{2.33}$$

$$\because \frac{a}{2r} << 1, \qquad \therefore V = \frac{e\,a\cos\theta}{r^2} \tag{2.34}$$

where θ is the angle between the axis of the dipole and the direction of r. The components of the electrical field (E) in the radial and transverse directions can further be obtained:

$$E_r = -\frac{\partial V}{\partial r} = \frac{2e\,a\cos\theta}{r^3} \tag{2.35}$$

$$E_\theta = -\frac{\partial V}{r\partial\theta} = \frac{e\,a\sin\theta}{r^3} \tag{2.36}$$

The quantity ea is usually called the dipole moment (μ), defined as the product of one of the dipole charges and the distance between the dipole charges. Consequently, the test charge present at P is exerted by a force (F) from the dipole. Since $F = eE$, and combining Eqs. (2.35) and (2.36), we obtain

$$F_r = \frac{2\mu e\cos\theta}{r^3}, \qquad F_\theta = \frac{\mu e\sin\theta}{r^3} \tag{2.37}$$

2.1.2.4.2 Induced Dipoles

The dipoles induced by the external field are called induced dipoles. The external field may arise from an instantaneous dipole; for instance, the electrons, in moving around the nuclei, do not have centers of gravity that instantaneously coincide with the nuclei, and the system would become a dipole. The induced dipole moment is defined as

$$\mu_I = \alpha E \tag{2.38}$$

where α is the polarizability and E the electrical field.

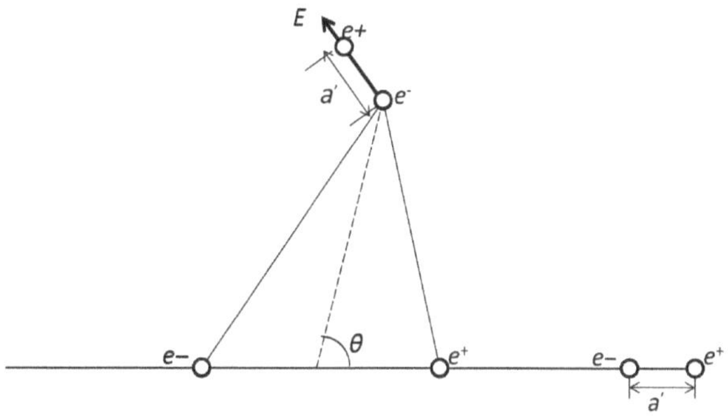

FIGURE 2.5 Induced dipole developed in the dipole–dipole interaction for a pair of inert-gas atoms.

Taking into account two inert-gas atoms being brought close together, an instantaneous dipole may occur in one atom with dipole moment μ. Assume the other inert-gas atom locating at an angle θ with respect to the instantaneous dipole, as represented in Fig. 2.4. The electrical field generated by the instantaneous dipole would make the other inert-gas atom become an induced dipole with distance a' along the electrical field, as shown in Fig. 2.5.

The variation of the electric field intensity with the distance (r) along the direction of the electrical field would be a function of $1/r^3$, as per Eqs. (2.35) and (2.36). Let the position of the negative charge of the induced dipole be r with the field intensity (E) whose magnitude is

$$E = (E_r^2 + E_\theta^2)^{1/2} = \left(\frac{4\mu^2\cos^2\theta}{r^6} + \frac{\mu^2\sin^2\theta}{r^6}\right)^{1/2} = \frac{\mu}{r^3}(3\cos^2\theta + 1)^{1/2} \quad (2.39)$$

The corresponding field at the positive charge of the induced dipole locating at the distance $r + a'$ is $E - \Delta E$. Since a' is very small, the corresponding curve can be assumed to be linear, and thus $\Delta E = \left|\frac{dE}{dr}\right| a'$, where $\left|\frac{dE}{dr}\right|$ is an absolute value of the slope. Thus the total force on the induced dipole due to the field of the instantaneous dipole is:

$$\begin{aligned} f &= -e'E + e'\left(E - \left|\frac{dE}{dr}\right| a'\right) \\ &= -e'a'\left|\frac{dE}{dr}\right| = -\mu_I\left|\frac{dE}{dr}\right| \\ f &= -\alpha E\left|\frac{dE}{dr}\right|, \end{aligned} \quad (2.40)$$

Substitute Eq. (2.39) in Eq. (2.40), yielding

$$
\begin{aligned}
f &= -\alpha\frac{\mu}{r^3}(3\cos^2\theta+1)^{1/2}\left|\frac{d}{dr}\left(\frac{\mu}{r^3}(3\cos^2\theta+1)^{1/2}\right)\right| \\
&= -\alpha\frac{\mu}{r^3}(3\cos^2\theta+1)^{1/2}\left(3\frac{\mu}{r^4}(3\cos^2\theta+1)^{1/2}\right) = -\frac{3\alpha\mu^2}{r^7}(3\cos^2\theta+1)
\end{aligned}
\tag{2.41}
$$

At $\theta = 0$, i.e., instantaneous and induced dipoles are at the same line (see Fig. 2.5)

$$
f \approx -\frac{\alpha'\mu^2}{r^7} \tag{2.42}
$$

where $\alpha' = 12\alpha$. The energy of the dipole-induced dipole interactions can be further calculated as follows:

$$
\begin{aligned}
f &= -\frac{\partial\phi}{\partial r} \\
\phi &\cong \int_{\infty}^{r} \alpha'\frac{\mu^2}{r^7}dr \cong -2\alpha\frac{\mu^2}{r^6}
\end{aligned}
\tag{2.43}
$$

For a solid, the van der Waals attractive energy can be considered to be the summation of the dipole interactions arisen from a synchronization of the motion of the electrons on the various atoms throughout the crystal. Nevertheless, the obtained energy still varies as the inverse sixth power of the distance as represented in Eq. (2.43); whereas it would be more accurately corrected by modern quantum mechanical theory expressed in terms of a single ion:

$$
\phi \cong -\left(\frac{C_1}{r^6} + \frac{C_2}{r^8} + \frac{C_3}{r^{10}}\right)
$$

where the first term represents the interactions of dipole–dipole; the second term dipole–quadrupole (16% of the total energy contribution); and the third term quadrupole–quadrupole (1.3% of the total energy contribution).

2.1.2.5 Zero-Point Energy

An important aspect of quantum theory is the existence of zero-point energy, that is, localized particles can never be completely stationary and always possess some kinetic energy. Zero-point energy is fundamentally related to the Heisenberg uncertainty principle. Briefly speaking, the lowest-energy state (the ground state) of the system must have a distribution in position and momentum that satisfies the uncertainty principle, which implies the energy of the ground state must be greater than the minimum of the potential well of the harmonic crystal. Thus the zero-point energy of a crystal can be considered to be the "vibrational energy" of the atoms in the ground state. However, this ground state energy cannot be shared with other atoms and does not count as heat or disorderliness. Moreover, the motion of oscillations can be broken into modes. Einstein treated each atom independently and also assumed that each oscillation has the same frequency, and so the number of modes is equal to the number of atoms. However, Debye considered that each mode has its own frequency or k (wave vector) value, which is more reasonable and will be taken into account. The assumptions of the model made by Debye are as follows:

1. The vibrations of a body, regarded as a continuous elastic solid, are considered as a whole, where the force of interaction is equivalent to a linear spring.
2. The lattice vibration of a solid can be viewed as a phonon gas. It is further assumed that simple sine waves of certain frequencies and amplitudes can be generated by hypothetical oscillators throughout the crystal, where the amount of the displacement of the atoms away from their equilibrium positions is equal to the amplitude of the sine wave at that point. Thus the superposition of such waves will simulate the complicated pattern of the actual atomic vibrations.
3. In a vibrating solid, there are three types of waves, i.e., one longitudinal and two transverse waves. However, longitudinal lattice vibration is ignored, and the vibration modes of such an atom array are analogous to the standing wave (transverse mode).

Since each oscillator of the assembly vibrates with its own frequency, and an assembly of 3N linear oscillators is taken into consideration, there must be an upper limit to the frequency. Considering a solid to be a periodic array of mass points, there are constraints on both the minimum and maximum wavelength associated with a vibrational mode. The minimum wavelength or the mode of maximum frequency is obtained when neighboring atoms vibrate against each other, namely, $\lambda_{\min} = 2a$ (where a is interatomic distance or lattice constant).

The maximum vibrational frequency in solids can simply be obtained:

$$\nu_m = \frac{v}{\lambda_{\min}} = \frac{v}{2a}$$

Taking the velocity as that of the shortest sound waves, $v = 5 \times 10^3\ m/s$ and assuming $a = 0.25$ nm, we obtain

$$\nu_m = \frac{5\times 10^3}{2\times 0.25\times 10^{-9}} = 10^{13}\ \text{Hz}$$

This value is often called natural frequency of the atom vibration in solids, usually used for the unknown jumping frequency in solid-state diffusion.

Consider the atoms as the quantum harmonic oscillators whose energy states are quantized; according to the quantum theory, its energy values can be represented as $(n_i + 1/2)h\nu$, where n_i is the quantum number associated with the ith oscillator and the $1/2h\nu$ accounts for the zero-point energy of each atom. The zero-point energy of a solid (E_z) can be obtained by summing over all the atoms:

$$E_z = \int_0^{\nu_m} \frac{1}{2} h\nu f(\nu) d\nu \tag{2.44}$$

where $f(\nu)$ is the Debye density function expressed as

$$f(\nu) = \frac{9N\nu^2}{{\nu_m}^3} \tag{2.45}$$

Combining Eqs. (2.44) and (2.45), gives

$$E_z = \frac{9}{8} Nh\nu_m \tag{2.46}$$

where ν_m is the cut-off frequency, which can further be represented as $\nu_m = \left(\frac{3N}{4\pi V}\right)^{1/3} c$ (where c is the velocity of sound and V (volume) $= a^3$).

Finally, replace Eq. (2.11) concerning Born's repulsive energy with a more precise expression:

$$V = B\cdot \exp\left(-\frac{r}{\rho}\right) \tag{2.47}$$

where ρ : constant ≈ 0.35; and thus the lattice energy can be represented by a more complete equation, combining Eqs. (2.8), (2.47), (2.43), and (2.46), for one mole of crystal:

$$U_L = -\frac{A\cdot e^2\cdot N}{r} + BNe^{-\frac{r}{\rho}} - CNr^{-6} + (9/8)Nh\nu_m \tag{2.48}$$

Furthermore, the lattice energy can be evaluated by the following empirical method called the Born–Haber cycle.

2.1.2.6 *The Born–Haber Cycle*

Born and Haber applied the Hess' law to ionic crystals, devising the so-called Born–Haber cycle, and allowing the enthalpy or heat of formation of an ionic crystal to be evaluated. This can be further applied to the lattice energy of an ionic crystal since it is equivalent to its heat of formation from one mole of its ionic constituents in the gas phase. We will set MX as an example as follows. Basically, the following reaction cannot be measured experimentally,

$$M^{+}_{(g)} + X^{-}_{(g)} \rightarrow MX_{(s)} \quad \Delta H = U_L \text{ (exo)} \tag{2.49}$$

Hence, the heat of formation of a crystal (ΔH_f) (Eq. (2.50)) instead can be calculated by constructing a thermochemical cycle, i.e., the Born–Haber cycle.

$$M_{(s)} + \frac{1}{2}X_{2(g)} \xrightarrow{\Delta H_f} MX_{(s)} \tag{2.50}$$

On the basis of Eq. (2.50), the Born–Haber cycle can be constructed by the following hypothetical reaction pathways:

Sublimation of $M_{(s)} \rightarrow M_{(g)}$	$\Delta H = S$(endo)
Dissociation of $1/2X_{2(g)} \rightarrow X_{(g)}$	$\Delta H = 1/2D$(endo)
Ionization of $M_{(g)} \rightarrow M^{+}_{(g)}$	$\Delta H =$ IE(endo)
Formation of X^{-} ion,	$\Delta H =$ EA(endo)

Coalescence of gaseous ions to give crystalline MX (Eq. (2.49)), $\Delta H = U_L$

According to the first law of thermodynamics or Hess' law, the heat of formation energy is given by the summation of the above energy terms generated by each hypothetical reaction pathway. Thus

$$\Delta H_f = S + 1/2D + \text{IE} + \text{EA} + U_L \tag{2.51}$$

Take MX as NaCl whose energy values are *S*: 109, 1/2*D*: 121, IE: 493.7, EA: −356, and U_L: −764.4 kJ mol^{-1}. According to Eq. (2.51), the calculated value of ΔH_f is −396.7 kJ mol^{-1}, reasonably agreeing with the measured value of −410.9 kJ mol^{-1}. This result confirms that the simple model for the evaluation of the lattice energy of an ionic solid is, for the most part, correct.

2.2 COVALENT BONDS

As mentioned in Section 2.1, a crystal with a difference of the EN lower than 1.7 is considered to be predominated by covalent bonds, indicating that the valence electrons are sharing rather than tightly bound to the respective atoms inside the crystal.

2.2.1 Formation Energy and Mechanism of Covalent Bonding

The bond energy of covalent bonding can be understood via how the covalent bond develops by an elementary consideration of the hydrogen molecule whose lowest atomic energy state is associated with 1s shell (Rioux, 2001). According to the Pauli exclusion principle, two electrons with parallel spin are unlikely due to the increase of the system energy, leading to an unstable molecule. As for two electrons with opposite spin, there are two possibilities for these two electrons to be distributed between these two nuclei. One possibility is that these two electrons spend more time in the neighborhood of one nucleus than the other, forming an ionic structure, but the formation energy is high, around −1237 KJ/mol, which is unstable, contributing 5% to the total binding energy. Another possibility is relevant to electron interchange, the so-called resonance effect, where both electrons simultaneously exchange nuclei, occurring at a very rapid rate, contributing about 80% to the total binding energy. The final energy concerning the electrostatic interactions, i.e., the interactions between the two electrons and the two protons contribute 15% to the total energy.

In most textbooks, the formation of covalent bonds is mainly attributed to that on the average, the electrons spend more of their time in the region between the two protons than they do on the far sides of the protons. However, it has been pointed out that this kind of interpretation is incorrect on at least two counts (Bacskay et al., 1997). First, it completely neglects the significance of electronic kinetic energy and second, the accumulated charge in the bonding region is so small that any increased attraction between the nuclei and electrons as a consequence of such charge accumulation cannot overcome the nuclear repulsion; thus no strong bonding would be predicted. The mechanism of the formation of a covalent bond can be interpreted in terms of two phenomena concerning orbital contraction and charge delocalization. As the molecule forms, the atoms are promoted by orbital contraction, which is atomic and endothermic. Because the electrons move closer on average to their respective nuclei, the potential energy decreases. By contrast, the kinetic energy increases because of the greater confinement of the electrons in the contracted orbitals, due to the fact that kinetic energy is inversely proportional to the square of the average distance of the electron from the nucleus. In this step, the increase of the kinetic energy is more than the decrease of the potential energy.

The constructive interference accompanying the overlap of the promoted atomic orbitals would bring about charge delocalization and charge redistribution. Charge delocalization indicates that the electron density distributes over the molecule as a whole, i.e., each electron now belongs to both nuclei, resulting in a significant decrease in kinetic energy. However, charge redistribution would transfer some electron density from the neighborhood of the nuclei into the internuclear region, leading to the increase in electron potential energy. This step is exothermic because kinetic energy decreases more than potential energy increases.

The orbital contraction does decrease the total potential energy of a molecule as its constituent atoms move towards each other, which increases the localization of the electrons around the nuclei and, incidentally, decreases the accumulated charge in the bonding region(s). The abovementioned mechanism clearly shows that covalent bond formation in H_2 is driven by a decrease in kinetic energy brought about by the charge delocalization that accompanies the overlap of atomic orbitals.

Generally, covalently bonded solids needed for atomic stability have to complete the octet of electrons following the so-called (8-*N*) rule, where *N* is the number of valence electrons in the given element, and the factor (8-*N*) indicates the number of the nearest neighbor in the structure. Nevertheless, to form a solid, each atom has to be simultaneously bonded to at least two other atoms, wherein hybridization between the s and p orbitals would be necessitated.

2.3 METALLIC BONDS

The valence electrons inside the metals are assumed to be nearly-free. Metallic bonds arise from the electrostatic attractive force between these nearly-free valence electrons (delocalized electrons) in the form of an electron cloud and positively charged metal ions (cores). It should be noted that a metallic bond is not the only type of chemical bonding that a metal can exhibit, even as a pure substance. For example, elemental gallium consists of covalently-bound pairs of atoms in both liquid and solid state—these pairs form a crystal lattice with metallic bonding between them. Another example of a metal–metal covalent bond is the mercurous ion (Hg^{2+}).

2.3.1 Cohesive Energy of Metals

Major contributions to the total energy of cohesive energy of metal can be deduced from the interactions between the electron cloud and the core in the following:

1. The Coulomb repulsive energy between positive ion cores.
2. The energies associated with the valence electrons arising from the following consideration including (a) the kinetic energy; (b) the potential energy in the field of the ion cores; (c) the coulomb self-potential; (d) the exchange interaction energy between electrons of parallel spins; and (e) the correction to (d) for electrons of antiparallel spin.
3. The exchange and correlation interactions between valence and core electrons.

It should be noted that the repulsive energy in most metals becomes large as the distance between nuclei is made smaller and is chiefly due to the electronic kinetic energy term. It is also noted that the important difference between metallic and covalent materials is the directionality of bond of the latter.

2.4 EFFECT OF BONDING ON THE MATERIAL PROPERTIES BY MEANS OF THE POTENTIAL-WELL CONCEPT

Many approaches like the Lennard–Jones for rare-gas solids, the nearly free electron for metals, and formation of H_2 molecule for covalent solids have been developed in efforts to describe the crystal potential energy. Moreover, the first-principles quantum mechanical calculations of the crystal potential energy as a function of the lattice spacing have been performed. Each of the calculations gave a curve similar to that shown in Fig. 2.3 for ionic crystals except the formation of ions from neutral atoms. It should be noticed that such a curve is referring to the static or nonvibrating crystal. The exact shape of the potential energy well is crucial for determining the material properties.

2.4.1 Elastic Modulus

Elastic modulus is a measure of the stiffness of a material and the theoretical strength of that material, which can be understood by using the potential-well concept. While the atomic viewpoint concerning the relationship between theoretical stress and elastic modulus has been obtained (Eq. (2.24)), this section presents a somewhat macroscopic view regarding the relation between elastic modulus and shapes of the potential well. When an external force is applied to a material, the force tends to push the atoms together under compression at a new position $r < r_0$ or pull the atoms apart under tension at $r > r0$, where r_0 is the equilibrium atomic position. At the new position, the sum of the atomic and applied forces acting on the atoms is zero. Let $r - r_0 = \Delta r$. Since Δr is very small, the potential energy of the displaced atom can be expanded as a Taylor series:

$$V(\Delta r) = V_{r_0} + \left(\frac{dV}{dr}\right)_{r_0} \Delta r + \frac{1}{2}\left(\frac{d^2V}{dr^2}\right)_{r_0} \Delta r^2 + \text{higher order terms} \quad (2.52)$$

Higher order terms could be neglected because of the small Δr. In addition, since

$$\frac{dV}{dr} = 0 \text{ at } r = r_0,$$

$$V(\Delta r) = V_{r_0} + \frac{1}{2}\left(\frac{d^2V}{dr^2}\right)_{r_0} r^2 \tag{2.53}$$

The atomic force, F, is given by:

$$F = -\frac{dV(\Delta r)}{d(\Delta r)} = -\left(\frac{d^2V}{dr^2}\right)_{r_0} \Delta r \tag{2.54}$$

Eq. (2.54) shows the linearity between the force and the displacement Δr, in which the constant of proportionality is related to the curvature of the potential energy well at its minimum. The proportionality constant is called the elastic modulus if we express force and displacement in terms of stress and strain. Since the curvature of the shape of the potential well determines the modulus, a narrow steep potential energy will correspond to a high modulus, whereas a broad, shallow energy represents a low modulus. It should be mentioned that elastic modulus is the same for both tension and compression.

2.4.2 Melting Temperatures

At the melting point, transformation would occur from the solid to the liquid state, wherein the amount of heat known as the heat of fusion ΔH_f would be absorbed, in other words, it is a measure of the enthalpy difference between solid and liquid states at the melting temperature. With an atomic viewpoint, melt would occur when the bond or lattice energy of atoms in a crystal is overcome by sufficient thermal energy. Referring to Fig. 2.3, the bond or lattice energy is determined mainly by the depth of the energy well, and so is the melting temperature.

2.4.3 Thermal Property

As far as the thermal expansion is concerned, the vibrations of the atoms should be taken into account. The harmonic crystal, wherein the atoms are considered to be connected by ideal springs and vibrate in harmonic motion, is adopted. Underlying the classical theory concerning the harmonic crystal, taking the equilibrium interatomic distance as a reference point, the potential energy as a function of the amplitude is parabolic, shown in Fig. 2.6A. A_i ($i = 1, 2, 3$, etc.) represents the turning point, where the potential energy is the maximum, but the kinetic energy is zero. The total energy of the position between two turning points includes the potential energy and kinetic energy at that position. For the quantum harmonic crystal, the derived energy states are discrete with the lowest energy, called the zero-point energy, where the probability density function indicates that the most probable position is at the zero amplitude, implying that the vibration amplitude is extremely small

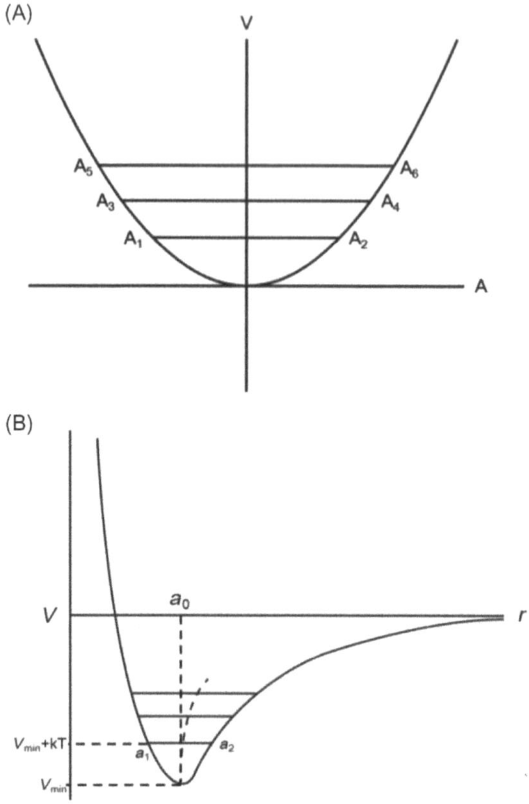

FIGURE 2.6 (A) The potential energy as a function of the amplitude of the atoms in the classical harmonic crystal, where A_i ($i = 1, 2, 3$, etc.) are the turning points. (B) The potential energy as a function of the interatomic distance of the static or nonvibrating crystal. At the absolute zero, it is assumed that all of the atoms have the lowest possible energy, $V_{\min}$ with the equilibrium separation a_o. At temperature T, the energy is raised to $V_{\min} + \mathrm{kT}$, where a_1 and a_2 correspond to the turning points in (A). The mean positions at different temperatures are at a separation greater than a_o, indicating the thermal expansion.

at this state. As the temperature is increased, more probability would occur at the turning points. At very high temperature, the probability density function reveals that the vibration behavior of the quantum harmonic crystal will approach to that described by the classical theory.

It is well known that thermal expansion is a result of the anharmonicity of the vibration. Hence, if the potential energy curve were symmetric, there would be no net change in interatomic separation with rise of temperature and, consequently, no thermal expansion. Apparently, a purely harmonic crystal potential will not result in thermal expansion. In the light of both curves of the static (or nonvibrating) (Figs. 2.3 or 2.6B) and the harmonic crystals (Fig. 2.6A) basically describing internal or lattice energy in terms of the interatomic spacing, most people resort to the former curve in interpreting the thermal expansion of materials. However, this curve per se is static and merely considers the interatomic distance at the turning points of the atom vibration. Thus it is not proper to rationalize the thermal expansion with the atoms moving to and fro between the well of Figs. 2.3 or 2.6B. In addition, at absolute zero, the amplitude of the classical harmonic oscillators is zero, and as for the quantum harmonic oscillators, it is in a ground state with zero-point energy whose amplitude is extremely small. Hence, it is reasonable to assume that the atoms are in equilibrium positions, which in turn correspond to the minimum point of Figs. 2.3 or 2.6B. On the basis of this assumption, we can start to explicate the thermal expansion, resorting to Fig. 2.6B, in which at absolute zero, we assume that all atoms have the lowest possible energy, V_{min}, and the mean interatomic spacing corresponds to the equilibrium separation a_{o}. At a temperature T, the average energy is increased by kT, where a_1 and a_2 correspond to the turning points A_i in Fig. 2.6A. In addition, at each temperature, it is also rationalized that the amplitude of thermal vibrations for an atom varies with the width of the potential-well curve, and the average interatomic distance is represented by the mean position. As the result of the asymmetry of the potential well, the mean positions with rising temperatures are greater than a_o, shown in Fig. 2.6B, which causes the positive thermal expansion. Since the amplitude of the vibration is dictated by the shape of the potential energy well, for a deep narrow energy well, the change in interatomic distance is small with a low thermal expansion coefficient, which is observed in materials having strong bonding energies, but for a broad shallow energy well, the change in interatomic distance is large with high thermal expansion coefficient, as observed in material having small bond energies.

Thermal expansion can be characterized by the linear thermal expansion coefficient (α), defined as the fractional change in length with the variation in temperature at constant pressure. Note that it is important to specify the temperature range over which α is measured since α increases with increasing temperature. This reflects the fact that the energy well becomes more asymmetric as one move up the well, i.e., with increasing temperature.

Chapter 3

The Structures of Crystalline Crystals

From the viewpoint of thermodynamics, the equilibrium structural state of a system is considered to be the minimization of the free energy under some constrains, such as constant temperature, pressure, and composition. For a solid, the related free energy can be approximated as internal energy. Thus the free energy is minimized by minimizing internal energy with reference to the lattice energy that is relevant to the interatomic potentials. Since the interatomic potential usually decreases with decreasing separation distance, most materials would adopt a very densely packed structure at low temperature. Such low-temperature, densely packed states can be said to be "internal energy stabilized" or to have a "high degree of order."

At sufficiently high temperatures, entropy would become dominant. Thus if the system possesses high entropy, namely, it adopts less well-ordered atomic/molecular arrangements, the free energy would be minimized. The high-temperature, more disordered equilibrium states can be referred to be "entropically stabilized" or as having a "low degree of order." Note that for the metal iron, in addition to spatial locations of the iron atoms, the free energy depends on the structural order of magnetic spins, leading to the more open structure (BCC) at low temperatures.

3.1 ARRANGEMENTS OF ATOMS AND IONS IN CRYSTALLINE SOLIDS: SPACE LATTICE

The essential characteristic of the crystalline state is the regularity in the arrangement of atoms. While the ways to describe the crystal structures are considerable, the science of crystallography has established a convenient system of classifying these structures. The structure is defined as a periodic array of lattice points associated with the pattern possessing translation symmetry, in which the surroundings viewed from an arbitrary origin are identical to the surroundings viewed from a point separated from the origin by a translation vector. The set of lattice points construct a lattice. Thus the simple way to describe the crystalline structure is to consider that it consists of a three-dimensional arrangement of lattice points in space, shown in Fig. 3.1, where a, b, and c represent the unit length of each axis.

Elements of Structures and Defects of Crystalline Materials.
DOI: http://dx.doi.org/10.1016/B978-0-12-814268-4.00003-5

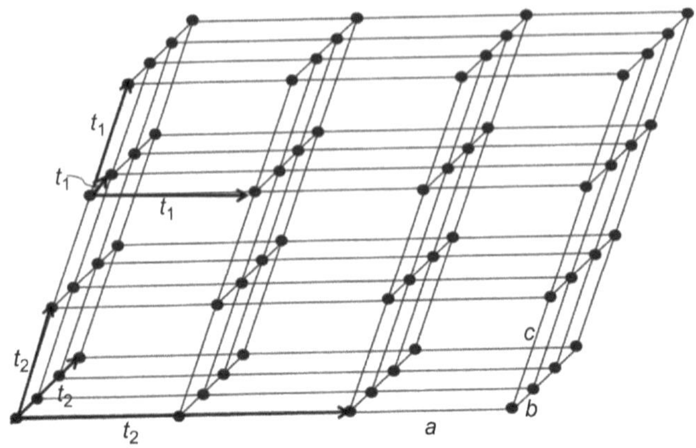

FIGURE 3.1 A three-dimensional arrangement of lattice points, indicating that each lattice point has identical surroundings, where a, b, and c represent the unit length of each axis and t_1 and t_2 represent translation vectors.

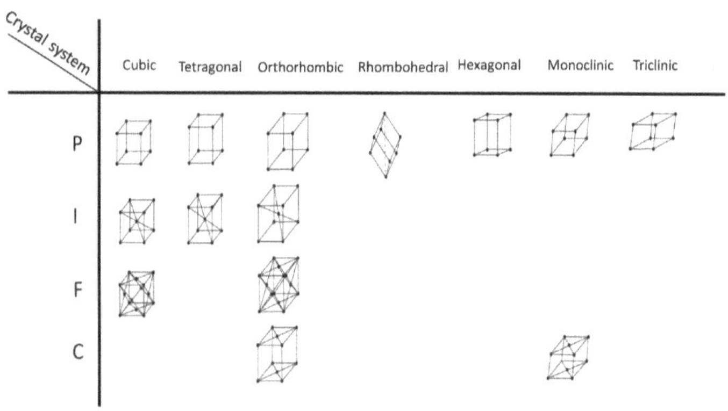

FIGURE 3.2 The 14 different Bravais lattices and seven crystal systems, where P is primitive cell, I is body-centered cell, F is face-centered cell, C is side-centered cell.

It was found that there are only 14 different ways to arrange the lattice points in space satisfying the definition of a space lattice, i.e., each point has identical surroundings. The 14 different point lattices are called Bravais lattices and are shown in Fig. 3.2.

In general, any lattice can be described in terms of nonparallel translation vectors (vectors connecting the lattice points), shown in Fig. 3.1. The unit cell, the repeating unit of space lattice, is defined by three lattice translation vectors. Thus the edge of unit cell must be lattice translation, and the magnitude of "lattice translation vectors" is called lattice parameter or lattice constant. While one space lattice can have a number of different unit cells, conventionally, unit cells are chosen which have a simple geometry and contain only a few lattice points. If three nonparallel edge vectors of the cell are chosen so that the only

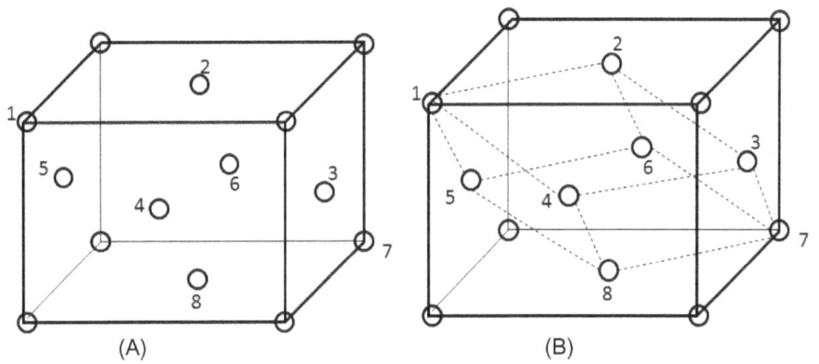

FIGURE 3.3 The unit cell (A) and the primitive cell (B) represented by dashed lines of an FCC lattice.

lattice points in the cell are at its corners (or equivalently, so that there is only one lattice point per cell), such a unit cell is of a unique kind and is called a primitive cell. As observed in Fig. 3.1, the cell with t_1 translation vectors is a primitive cell containing a single lattice point but the cell with t_2 translation vectors is a nonprimitive cell containing more than one lattice point. The choice of nonprimitive cells is due to the fact that it is mathematically more convenient to describe the crystal structures than their primitive cells. FCC lattice is set as an example, illustrated in Fig. 3.3, where the primitive cell is marked with dashed lines, and in order to make the primitive cell clear, the related lattice points are labeled with numerical values.

In addition, the nonprimitive cell can be developed in terms of the primitive cell by assigning each lattice point with a basis, defined as the atom group, which has an identical composition arrangement and orientation. The hexagonal system of Hexagonal closely packed (HCP) and nonprimitive hexagonal structures will be taken as the examples, illustrated in Fig. 3.4. Concerning the HCP structure, it can be developed by adopting the hexagonal primitive cell associated with a lattice point with a basis of (0, 0, 0) and (1/3, 2/3, 1/2), which then grafts into each primitive cell of a side-centered hexagonal cell. As a result, the HCP structure can be formed. Likewise, the nonprimitive hexagonal cell can be formed by taking the basis of (0, 0, 0), (2/3, 1/3, 1/3), and (1/3, 2/3, 2/3), also shown in Fig. 3.4, which can then further deduce the primitive rhombohedral Bravais lattices defined by "rhombohedral" basis vectors t_1', t_2' and t_3'. This is the reason for the adoption of six instead of seven crystal systems, listed in Fig. 3.2.

3.2 METALLIC STRUCTURES

As aforementioned, the metal bonding has very little preferred directionality. Except for four metals, metals form in one of the following structures: face-centered-cubic (FCC), body-centered cubic, hexagonal closely

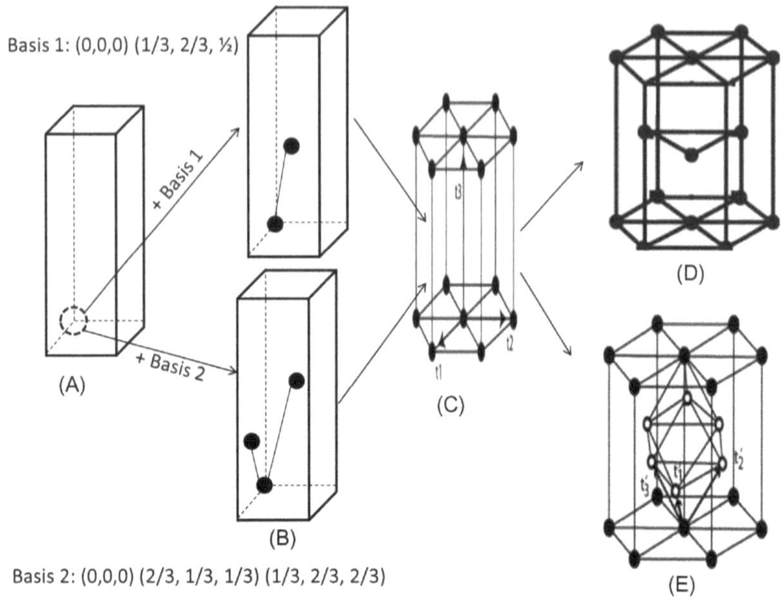

FIGURE 3.4 Schematic illustrations of the steps to develop HCP structure and the rhombohedral Bravais lattice from the primitive hexagonal lattice (A). At first, let a lattice point of the primitive hexagonal lattice be combined with basis 1 and 2 (B), grafted into a side-centered hexagonal cell, defined by lattice vectors t_1, t_2, and t_3, respectively. For basis 1, HCP structure (D) can be obtained, and for basis 2, a nonprimative hexagonal cell (E) can be produced, corresponding to the rhombohedral Bravais lattice, which alternatively can be defined with basis vectors t_1', t_2', and t_3' to become a primitive rhombohedral cell.

packed, and tetragonal among the 14 Bravais lattices. The closely packed structures have been shown in Fig. 3.3 (FCC) and Fig. 3.4 (HCP), and their void locations in the stacking can be found in many textbooks.

3.3 IONIC STRUCTURES

The ions in ionic structures are regarded as a charged and polarizable sphere, which are held together by electrostatic force, and therefore, are arranged so that cations are surrounded by anions and vice versa. In order to maximize the net electrostatic attraction between ions in the structure (i.e., the lattice energy reduces), their coordination numbers are as high as possible. Moreover, next nearest neighbor interactions are of the anion–anion and cation–cation types and are repulsive, leading to a structure of high symmetry with a maximized volume. Nevertheless, local electroneutrality would be preserved, i.e., the valence of anions equal to the sum of the electrostatic bond strengths between them and adjacent ions of opposite charge.

3.3.1 Pauling's Rules

These rules predict the coordination of ions in ionic structures, described in the following.

3.3.1.1 Pauling's First Rule

This rule stated that a coordinated polyhedron of anions is formed about each cation, the cation–anion distance being determined by the radius sum and the coordination number of the cation by the radius ratio, which essentially introduces the concept of atomic (or rather ionic) size. This rule basically describes the geometric relationships between cations and anions in a stable structure, which infers that each cation is coordinated by a polyhedron of anions; the number of anions is determined by the relative sizes of the cation and anion, called the radius ratio rule wherein it assumes that a cation must be in contact with its anionic neighbors and this places a lower limit on the size of cation which may occupy a particular size, and neighboring anions may or may not lie in contact. Based on these assumptions, the radius ratios of anions and cations in different interstitial sites of an anion polyhedral can be evaluated. The octahedral site in a rocksalt structure (Fig. 2.2) will be taken as an example, where it is assumed that octahedral cations have a minimum radius r_m.

According to the geometric relationship shown in Fig. 3.5, we obtain:

$$(2r_x)^2 + (2r_x)^2 = [2(r_m + r_x)]^2$$

and therefore, $r_m/r_x = 0.414$.

Note that radius ratios would vary with the table of ionic radii selected. For instant, for RbI, $r^+/r^- = 0.69$, as taking $r_{o^{2-}} = 1.40$ but $r^+/r^- = 0.82$, as taking $r_{o^{2-}} = 1.26$. Moreover, when the radius of cation are larger than that of anion, e.g., for CeO_2, $r^+/r^- > 1$, the inverse ratio r^-/r^+ should be taken.

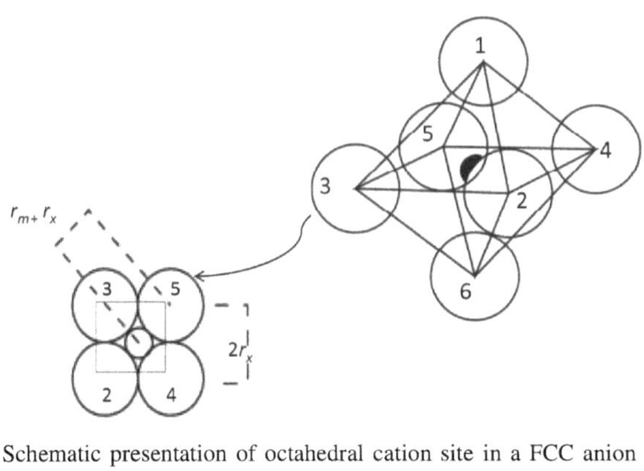

FIGURE 3.5 Schematic presentation of octahedral cation site in a FCC anion sublattice, indicating the geometric relationship between cation and anion.

Furthermore, it should also have to be pointed out that for materials with very complex compositions, its cation–anion distance does not depend only on the cation and anion involved, but also on the average radius of each of the other cations. Under this circumstance, a cation whose ionic radius is the largest but valence (formal charge) the lowest is the first to occupy the higher energy coordination, followed by smaller higher-valence cations which tend to form more rigid coordinations. For instance, in $Na_8Cl_{12}[Si_6Al_6O_{24}]$, Na^+ would locate at the tetrahedral site with CN being 4 (coordinated with $3O^{2-}$ and $1Cl^-$) rather than 6 as it usually is, due to the lowest valence.

3.3.1.2 Pauling's Second Rule (Electrostatic Valence Rule)

This rule states that in a stable coordination structure the electric charge of each ion tends to compensate for the strength of the electrostatic valence bonds, where the cation is at the center of a polyhedron with anions at the corners, and vice versa. This rule ensures that the basic coordination polyhedra are arranged in three dimensions in a way that preserves local charge neutrality. That is, in a stable ionic structure the valence of an anion must be balanced by an equal and opposite charge on the surrounding cations. However, because these cations are also shared by other anions, it is necessary to evaluate the effective charge associated with each cation–anion bond, called the electrostatic bond strength (e.b.s.) defined as the ratio of the charge on a cation to its coordination number. For a cation M^{m+} surrounded by n anions, X^{x-},

$$\text{e.b.s.} = \frac{m}{n} \tag{3.1}$$

According to this rule, for each anion, the sum of the electrostatic bond strengths of the surrounding cations must balance the negative charge on the anion, i.e., the valence of each anion x:

$$x = \sum_{\text{CN}} \frac{m}{n} \tag{3.2}$$

Taking spinel $MgAl_2O_4$ as an example, with Mg^{2+} located at tetrahedral sites with CN = 4, one obtains e.b.s. $= \frac{2}{4} = \frac{1}{2}$, but with Al^{3+} at octahedral sites, e.b.s. $= \frac{3}{6}$. Since the coordinates of oxygen are 3Al and 1Mg, $\sum \text{e.b.s.} = (3Al^{3+} + 1Mg^{2+}) = 2$. This result fulfills the Pauling's second rule. In the case where the coordinates of the anion are unknown and have to be determined, Pauling's second rule is insufficient to solve the problem, hence, one more criterion should be used to choose the correct one. The criterion is that the ratio of the number of cations bonded to oxygen/number of oxygen bonded to cation should be consistent with that of the values of cation and oxygen in the chemical formula. For instance, 3 for Al bonded to O/6 for O bonded to Al = 2/4.

Using the same argument, for rutile TiO_2. Ti^{4+} ions locate at octahedral sites, so e.b.s. $= \frac{4}{6} = \frac{2}{3}$, and O^{2-} is surrounded by 3 Ti^{4+}, thus

$$\sum_{i=1}^{3} \frac{2}{3} = 2$$

This result also meets this rule. Furthermore, one can predict that a common vertice in silicate structures cannot be shared by three SiO_4 units. Since Si^{4+} in a tetrahedral site, e.b.s. $= \frac{4}{4} = 1$. Hence, for an oxygen bridging two SiO_4 units, $\sum$ e.b.s. $= 2$, which is acceptable, but for three units, $\sum$ e.b.s. $= 3$, which is unacceptable.

In addition, in a structure with irregular polyhedra like brookite (a polymorph of TiO_2) which is characterized by strongly distorted TiO_6 octahedra whose cation–anion bond lengths are not all the same or for crystalline, non-molecular inorganic materials, in which insufficient bonding electrons are available for each bond (regarded as partial bonds), the bond-valence approach would be better in describing the related bonds. Since the approach of the bond valence is to treat each bond as an individual, irregularities or distortions in coordination polyhedral can be taken into account. Moreover, bond valences can also extend Pauling's second rule to structures which are not necessarily ionic by replacing (1) the e.b.s. with the bond valence, and (2) the formal charge on the anion with the valence of that atom. (Valence being defined as the number of electrons taking part in bonding.) This leads to the valence-sum rule relating the valence, V_i of atom i to the bond valence, b_{ij} between atom i and neighboring atom j:

$$V_i = \sum_j b_{ij} \tag{3.3}$$

The correlations between deviations from Pauling's second rule and bond length variations in crystals have been intensively studied, and several quantitative relations between bond length and the strength of a bond have been developed. During the 1960s and early 1970s, the term "bond strength" was used, but was later changed to "bond valence" (Brown, 2009) to distinguish these values from Pauling bond strengths. In the early 1970s, several different forms of the (inverse) relation between bond valence and bond length were used. Brown and Shannon (1973) proposed the following equation (Eq. (3.4)) of the bond valence:

$$b_{ij} = b_0(R/R_0)^{-N} \tag{3.4}$$

where R is the bond length, which can be deduced from the structure, b_0 is a parameter usually set to have R/R_o ~ 1, and R_o and N are the bond-valence parameters. They derived their bond-valence parameters in three different ways using least-squares fitting to minimize deviations from the valence-sum rule. For the incident bond-valence sums around the cations, they varied R_o and N or fixed N and varied R_o, but for the incident bond-valence sums around the cations and the anions, they varied R_o and N only. Furthermore, rather than use Eq. (3.4), they found a good fit of the following function:

$$b_{ij}^{0} = (R/R_1)^{-N_1}. \tag{3.5}$$

where R_1 and N_1 are constants for a given row. In most cases, the value of N_1 obtained does not differ by more than 1 from the individual N's of the atoms in the row. This allows a single pair of parameters to be used for all the elements in the same row that have isoelectronic ionic cores. Thus one curve of the form can be used to calculate bond strengths, e.g., the bonds between oxygen and atoms of the second row in their group valence states, namely, Na^+, Mg^{2+}, Al^{3+}, Si^{4+}, P^{5+}, and S^{6+}, etc. The universal correlation curve for oxygen ion bonding to these ions can be represented with $R_1 = 1.622$ Å, $N_1 = 4.290$ using Eq. (3.5). The fact that such a fit can be made implies that the length of the M–O bond of unit strength (R_1) is the same (or nearly the same) for all ions with the same electron core.

Furthermore, the following equation (Eq. (3.6)) proposed by Brown and Altermatt (1985) was eventually accepted as the general form of the bond-valence–bond-length curve:

$$b_{ij} = \exp\left(\frac{R_0 - R_{ij}}{b}\right) \tag{3.6}$$

where b_{ij} is the experimental bond valence, R_{ij} the observed bond length between ions i and j, and R_0 and b are fitted bond-valence parameters. R_0, which represents the nominal length of a bond of unit valence, depends on the sizes of the atoms forming the bond. The value of b is found to lie between 0.3 Å and 0.6 Å; however, because of the limited range of experimental bond lengths, its precise value is not easy to determine. Thus a value of 0.37 Å is frequently assumed. Using Eq. (3.6), a good fit has been achieved in bond-valence–bond-length correlation for Ca–O bonds with $R_0 = 1.967$ Å and $b = 0.37$ Å. Note that N in Eq. (3.4) or b in Eq. (3.6) is the measure of the softness of the interaction between the two atoms.

3.3.1.3 Pauling's Third Rule

This rule states that the presence of shared edges, and particularly of shared faces, in a coordinated structure decreases its stability; this effect is large for cations with large valence and small coordination number, and is especially large in cases where the radius ratio approaches the lower limit of stability of the polyhedron. It is inferred that coordination polyhedra prefer linkage where they share corners rather than edges, and edges rather than faces. This rule is simply based on the fact that cations prefer to maximize their distance

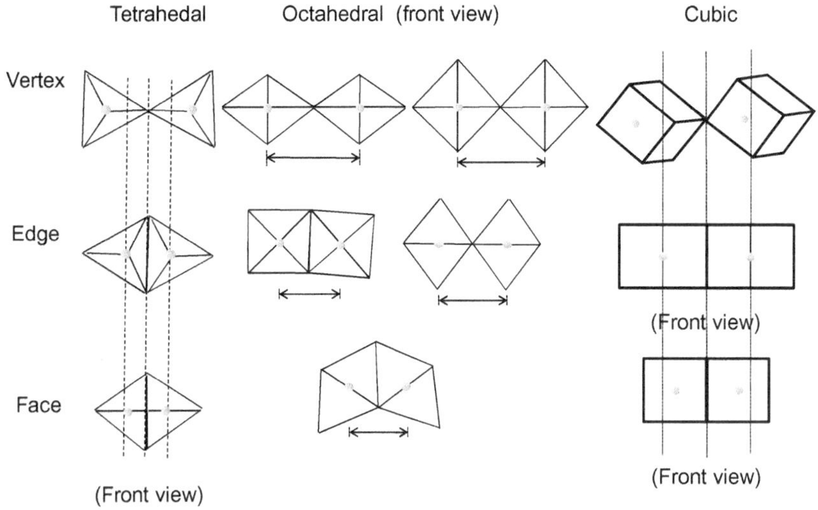

FIGURE 3.6 Schematic illustrations of the distances between cations of different polyhedral linkages.

from other cations in order to minimize electrostatic repulsive (with model polyhedra, one can visualize the decreasing distance between cations upon proceeding from corner-sharing to edge-sharing to face-sharing). The distances between cations of different polyhedral linkages have been schematically illustrated in Fig. 3.6.

Underlying the request of this rule, some principles in view of the geometrical relationships shown in Fig. 3.6 can be concluded:

1. The distances of tetrahedral coordinations sharing vertice, edge, and face are in the ratio 1:0.58:0.33, which make tetrahedral coordinations sharing an edge or a face tend to destabilize the structure due to the repulsion increase between cations, which explains why $[SiO_4]^{4-}$ is isolated or shares vertices.
2. Though the distances of octahedral coordination sharing edge and cubic coordination sharing face is the same (2.8 Å), the former would be more stable in view of the fact that the shielding of the two cations is more accentuated in this manner. This explains why numerous halites tend to keep the NaCl structure rather than that of CsCl.
3. On account of the fact that in cubic closely packed (CCP) structures the octahedral coordinations share edges while in HCP they share faces, an MX stoichiometry crystallizes not with an HCP, but with a CCP lattice (e.g., NaCl, MgO, CaO, etc.).
4. The relative stability of TiO_2 polymorphs is in the order of rutile, brookite, and anatase, depending on the number of the sharing edges of the octahedral coordination.

3.3.1.4 Pauling's Fourth Rule

This rule essentially is a corollary of third rule, and states that in a crystal containing different cations, those with large valence and small coordination number tend not to share polyhedron elements with each other. It is conspicuous that the absence of shared geometrical elements is attributed to the destabilizing effects of cation–cation Columbic repulsions in structures where they form, which also infers that highly charged cations with lower coordination number tend to be as far apart as possible to minimize the electrostatic repulsion. Thus compounds with cations in tetrahedral coordination polyhedra tend not to share edges or faces because structures would have distorted anion coordination environments which result in poor anion hybridization and weaker metal–anion bonds, rather, they tend especially to be linked by corner sharing, e.g., $[SiO_4]^{4-}$ is found either to be isolated or to share vertices.

Numerous structures violating the third and fourth rules, especially the third rule, are known. Conspicuous examples include β-BeO and the isotypic compounds $BeCl_2$ and SiS_2, all of which contain edge-sharing tetrahedra. Another striking example is the lovely dumortierite type which has crystal edge-sharing octahedral chains, face-sharing octahedral chains, and octahedral chains combining edge- and face-sharing.

3.3.1.5 Pauling's Fifth Rule (The Rule of Parsimony)

This rule (Burdett, 1984) stated that the number of essentially different constituents in a crystal tends to be small, wherein "constituents" refers to crystallographic sites (i.e., coordination polyhedra) rather than types of ions. It was also regarded as a corollary of the first rule, limiting variation in cation coordination number, and the second rule, limiting the number and type of cations around each anion. Owing to the fact that most ionic structures are closely packed by the anions with a variation with the smaller cations in the interstices, the number of the types of interstitial sites present within a periodically regular packing of anions tends to be small. Thus cations with a wide range of charges or sizes would tend to form nanoscale orderings or "superlattices," leading to an increase of the complexity of the structure. In addition, this rule is in favor of regularity, calling for high-symmetry arrangements of all ions, and it is certainly related to the more fundamental observation that most common materials are crystalline rather than amorphous.

3.4 STRUCTURAL DISTORTION IN IONIC STRUCTURES

The coordination number and environment of the metal atom in an ionic structure would be considerably influenced by two types of nonbonding electrons, i.e., d electrons in transition metal compounds and the lone-pairs S^2

electrons in compounds of the heavy p-block elements in low coordination states. Thus ionic crystals with these types of nonbonding electrons, d and the lone-pairs S^2 electrons, would yield structural distortion.

3.4.1 Effects of Crystal Field on the Structures of the Oxides

Crystal field theory is the fundamental for describing the structure development in several functional oxides. In crystal field theory, bonding is electrostatic deviation from the electric field of the ligands viewed as purely ionic species, whose concept is equivalent to that of the valence shell electron–pair repulsion theory.

3.4.1.1 Crystal Field in an Octahedral Coordination

In an octahedral coordinated polyhedron, the central ion has six ligands, resulting in the mutual repulsion of the six bonding electron pairs, which in turn would cause the splitting of the energy levels of the orbitals. For s and p orbitals, no splitting occurs because of symmetry of the orbitals, as observed in Fig. 3.7.

Regarding d orbitals, the five orbitals are degenerate (equal in energy) in a state of free ion. However, they would become differentiated, namely, split up into different energy levels, under the effect of the electrostatic field of the ligands. As the central ion in an octahedral coordinated environment, the electrons in the d_{z^2} and $d_{x^2-y^2}$ orbitals would encounter head-on interaction, therefore, they will exert greater repulsive force than those in d_{xy}, d_{yz}, and d_{xz} orbitals, which lie inbetween the metal–ligand bond axes and thus maintain a maximum separation from the ligands, as observed in Fig. 3.8.

Thus, the d orbitals of the central ion in an octahedral coordination are split into two groups: one has higher energy of the d_{z^2} and $d_{x^2-y^2}$ orbitals and the other has d_{xy}, d_{yz}, and d_{xz} orbitals with lower energy. It should be noted

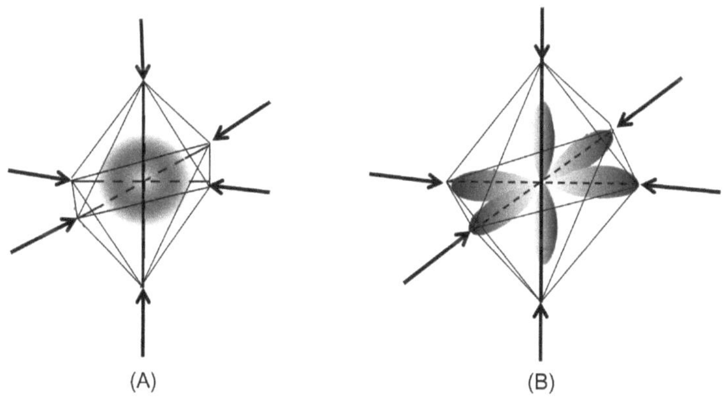

FIGURE 3.7 Crystal field in an octahedral coordination of (A) s orbitals and (B) p orbitals.

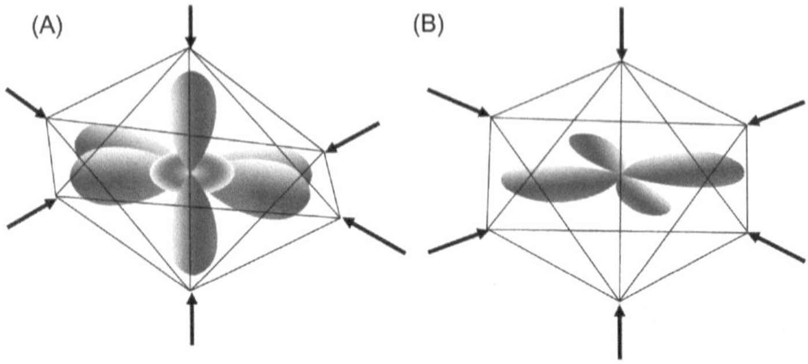

FIGURE 3.8 Crystal field in an octahedral coordination of (A) d_{z^2} and $d_{x^2-y^2}$ orbitals and (B) d_{xy} orbital.

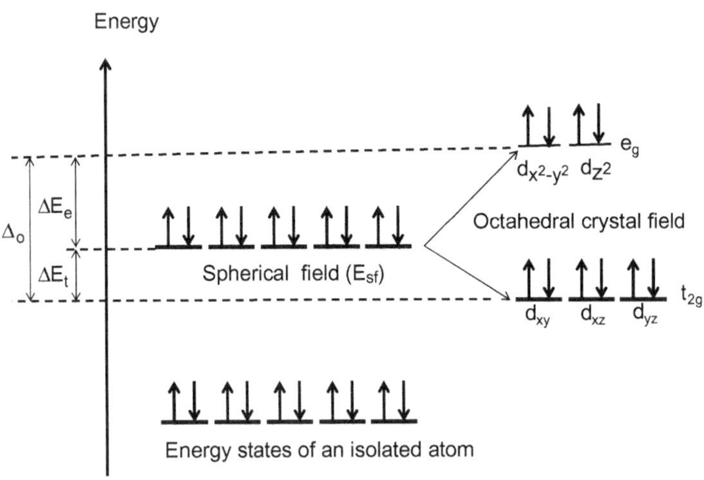

FIGURE 3.9 Energy levels of d orbitals in an isolated atom, spherical field, and octahedral field, where E_{sf} indicates the absolute energy of a d orbital of an ion in a spherical field.

that the energy levels of both groups are higher in comparison to those of the free ion, but it is customary to represent the splitting of the d orbitals relative to the ion in a spherical field, shown in Fig. 3.9.

As observed in Fig. 3.9, there are symbols specifying the characteristics of orbital symmetry of an ion in a crystal field. The related symbols are briefly reviewed in the following:

1. a or b indicates a nondegenerate orbital, and a is related to a wave function which has symmetry (+) with respect to the rotation axis, but b is with respect to a wave function which has antisymmetry (−) and changes sign during rotation.
2. e and t represent orbitals are doubly and triply symmetrical, respectively.

The following symbols are used in lower case notation associated with the above symbols:

g: the presence of a change in sign of the wave function or inversion through a center of symmetry;

1: the presence of mirror plane parallel to the symmetry axis;
2: mirror plane is normal to symmetry axis.

Based on the above definition of symbols, the d_{z^2} and $d_{x^2-y^2}$ orbitals are designated as e_g and the d_{xy}, d_{yz}, and d_{xz} orbitals are t_{2g}. Since the energy of the e_g orbitals is higher than that of t_{2g}, the electrons prefer the latter. The energy difference between e_g and t_{2g} orbitals is called crystal field splitting, represented as Δ_o, which is equal to 10 Dq.

The barycenter principle suggests that as the d orbitals are all filled, the sum of the energies of the splitting d orbitals is equal to that of the degenerated d orbitals of an ion in a spherical field. Underlying this principle, the energy lift of e_g levels indicated by ΔE_e and the lowering of t_{2g} levels indicated by ΔE_t in comparison to the levels of an ion in a spherical field can be evaluated. According to the conservation of energy, the energy of the splitting levels in an octahedral field should be the same as that of the degenerate levels in a spherical field or the raised and lowered energies should be the same. Taking fully-filled orbitals as an example, and as per Fig. 3.9, the following relations should be held:

$$4\Delta E_e = 6\Delta E_t \text{ and } \Delta E_e + \Delta E_t = \Delta_o, \tag{3.7}$$

In solving Eq. (3.7), we obtain $\Delta E_e = 0.6\,\Delta_o = 6\text{Dq}$ and $\Delta E_t = 0.4\Delta_o = 4\text{Dq}$. Note that the splitting of the energy levels Δ_o depends on the lobe of the central ion involving the shape of the electron cloud and the energy difference between occupied and unoccupied orbitals, and the ligand field, including the surrounding arrangement of the ligand field and the strength of the ligand fields.

The d electron configuration plays a significant role in determining structure distortion and material properties. Thus it is very important to analyze the distribution of the electrons in the splitting orbitals, wherein the Hund's rule and the pairing energy should be taken into consideration:

1. Hund's rule suggests that (1) every orbital in a sublevel is singly occupied before any orbital is doubly occupied and (2) all of the electrons in singly occupied orbitals have the same spin to maximize total spin. Actually, this is Hund's first rule referring to the maximum multiplicity that lies lowest in energy.
2. Pairing energy (P) results from the energy necessary to overcome the repulsion which exists between two electrons occupying the same orbital and to compensate the exchange energy which is lost when Hund's rule is violated.

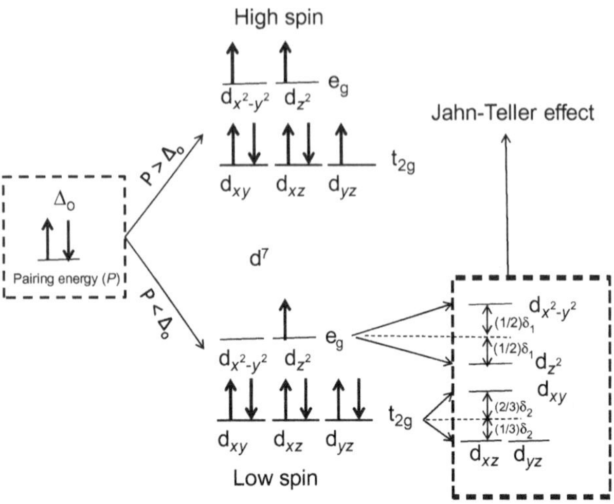

FIGURE 3.10 Schematic illustration of high-spin and low-spin states for a d^7 transition metal ion in an octahedral coordination environment as a consequence of the competition of pairing energy and crystal field splitting. The splitting of the energy levels for a d^7 ion, due to the Jahn–Teller distortion, is represented in the dashed box.

When the electrons are not more than three, they will occupy the t_{2g} orbitals, individually, according to Hund's rule. If there are more than three electrons, the extra electrons may have two alternatives, represented in Fig. 3.10 using d^7 as an example: if $P > \Delta_o$, the extra electrons would be more favorable to fill the e_g orbital first, but only two can occupy it according to Hund's rule, and the other two electrons remain at the t_{2g} orbital, forming a high-spin complex; if $P < \Delta_o$, the extra electrons would prefer to fill the t_{2g} orbital, and after the full filling, the remaining one electron occupies the e_g orbital, forming a low-spin complex.

3.4.1.2 Crystal Field in a Tetrahedral Coordination

Regarding the ions in a tetrahedral field, there are four ligands, as observed in Fig. 3.11.

It is apparent that the d_{xy}, d_{yz}, and d_{xz} orbitals, called t_2 orbitals, are closer to the ligand than the d_{z^2} and $d_{x^2-y^2}$ orbitals, called e orbitals. Thus the former orbitals would bear a larger electrical repulsion and have higher energy than the latter orbitals, contrary to the octahedral coordination. The value of the crystal field splitting is indicated as Δ_t. The splitting of d energy levels in a tetrahedral field is shown in Fig. 3.12.

Using the same argument as in the octahedral field, we obtain $\Delta E_e^{'} = 0.4\Delta_t$ and $\Delta E_t^{'} = 0.6\Delta_t$. In addition, $\Delta_t \approx \frac{4}{9}\Delta_o$. Δ_t is smaller than Δ_o due to the fact that there are four ligands instead of six and no direct (or

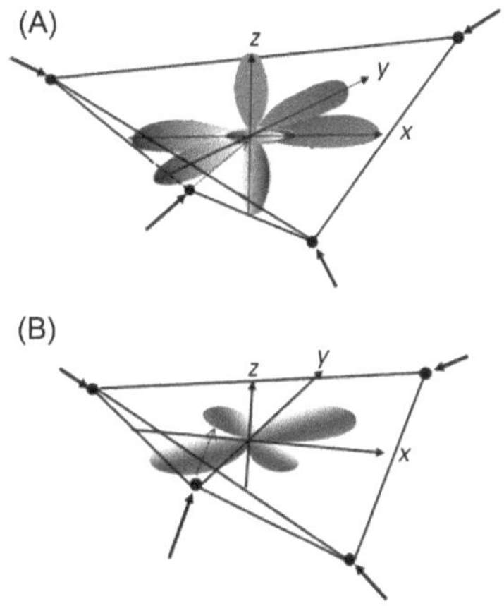

FIGURE 3.11 Crystal field in a tetrahedral coordination of (A) d_{z^2} and $d_{x^2-y^2}$ orbitals and (B) d_{xy}orbital.

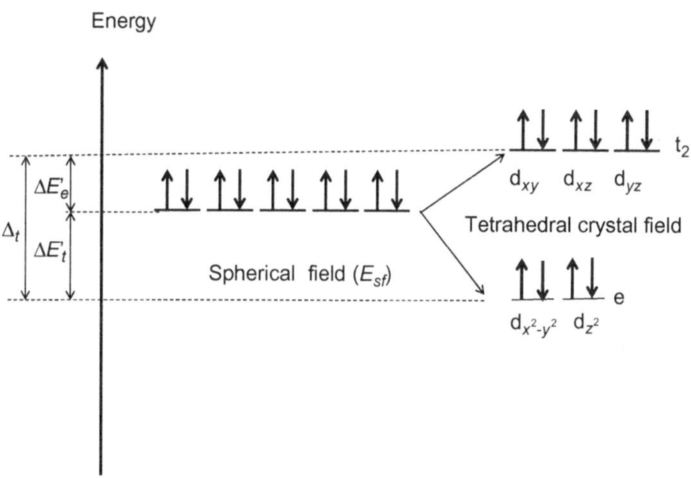

FIGURE 3.12 Energy levels of d orbitals in a spherical field and a tetrahedral field.

fewer) interactions between the ligand and the d orbitals of the metal ion. Because Δ_t is small, a tetrahedral field favors high-spin electronic configuration.

3.4.2 Crystal Field Stabilization Energy

In order to evaluate the structural option with the most favorable energetical state, crystal field stabilization energy (CFSE) is a useful criterion. Underlying the principle of the average weighted energy, the sum of the energies of the lift-up and the lowered states has to be equal to the energy of the spherical field. The value of CFSE in an octahedral field can be calculated from the algebraic sum $\sum_i x_i + nP$ of all the electrons of d orbitals, where i: ith electron; n: the number of paired electrons; P: pairing energy. As electrons are in lowered states, i.e., t_{2g} orbitals, $x_i = -0.4\Delta_o$ but in the raised states, namely, e_g orbitals, $x_i = +0.6\Delta_o$. For example, the values of the CFSE of the low and high spin states for Co^{+2} (d^7) in an octahedral field, referring to Fig. 3.10, are

$$\text{low spin: } (-0.4\Delta_o)\times 6 + 3P + 0.6\Delta_o \times 1 = -1.8\Delta_o + 3\text{P}$$

$$\text{high spin: } (-0.4\Delta_o)\times 5 + 2\text{P} + 0.6\Delta_o \times 2 = -0.8\Delta_o + 2\text{P}$$

Another factor useful for evaluating the site preference is octahedral site stability energy (OSSE), which is the value of the CFSE in octahedral sites minus that of the CFSE in tetrahedral sites. For most transition metal ions, their OSSE in octahedral sites is high, and thus they prefer octahedral coordination or a distorted variant of octahedral. The related values of the CFSE in octahedral with high spin and tetrahedral sites (in Δ_o units using $\Delta_t \approx \frac{4}{9}\Delta_o$) and OSSE are given in Table 3.1. The tendency between the calculated and experimental values (Dunitz, 1960) essentially is consistent, which shows

TABLE 3.1 The Calculated (in Parentheses) and Experimental (Obtained From Spectroscopic Data; Dunitz, 1960) Values of Crystal Field Stabilization Energies (CFSE) (kJ mol^{-1}) and Octahedral Site Stability Energy (OSSE) for Transition Metal Ions of the First Series in Oxides

Ions		CFSE (High Spin) Octahedral	CFSE (in Δ_oUnits) Tetrahedral	OSSE
Ti^{3+}	d^1	87.4(−0.4)	58.5(−0.27)	28.9(−0.13)
V^{3+}	d^2	160.1(−0.8)	106.6(−0.53)	53.5(−0.27)
Cr^{3+}	d^3	224.5(−1.2)	66.9(−0.35)	157.6(−0.85)
Mn^{3+}	d^4	135.4(−0.6)	40.1(−0.18)	95.3(−0.42)
Mn^{2+}	d^5	0(0)	0(0)	0(0)
Fe^{3+}	d^5	0(0)	0(0)	0(0)
Fe^{2+}	d^6	49.7(−0.4 + P)	33.0(−0.27 + P)	16.7(−0.13)
Co^{2+}	d^7	92.8(−0.8 + 2P)	61.9(−0.53 + 2P)	30.9(−0.27)
Ni^{2+}	d^8	122.1(−1.2 + 3P)	35.9(−0.35 + 3P)	86.2(−0.85)
Cu^{2+}	d^9	90.3(−0.6 + 4P)	26.8(−0.18 + 4P)	63.5(−0.42)
Zn^{2+}	d^{10}	(0 + 5P)	(0 + 5P)	(0)

that d^5 ions with high spin and d^0 and d^{10} ions have no particular preference for octahedral or tetrahedral sites as far as the crystal field effects are concerned. However, the ions of Cr^{3+}, Ni^{2+}, and Mn^{3+} with large OSSE reveal the strongest preference of the octahedral coordination.

According to the site preferences of ions, the spinel structure whose composition is specified as AB_2O_4 is classified as:

1. Normal spinel—A tetrahedral, B octahedral ($\gamma = 0$);
2. Inverse spinel—A octahedral, B tetrahedral and octahedral ($\gamma = 1$);
3. Some intermediate between normal and inverse ($\gamma = 0.67$), in which the parameter γ is defined as the fraction of A ions on octahedral site.

The cation distributions in the spinel structures are influenced by several factors. In the absence of CFSE, the relative sizes of A and B may in principle determine the spinel structure, in which the smaller cation prefers to occupy the site of lower coordination, i.e., the tetrahedral site. However, lattice energy may outweigh size factor, e.g., in $MgAl_2O_4$, Al^{3+} occupies the octahedral site, resulting in a normal spinel structure. On the basis of the lattice energy calculation, spinels of the type 2,3 (i.e., $A{=}M^{2+}$, $B{=}M^{3+}$, e.g., $MgAl_2O_4$) tend to be normal, whereas spinels of type 4,2 (i.e.,$A{=}M^{4+}$, $B{=}M^{2+}$, e.g.,$TiMg_2O_4$) tend to be inverse. As CFSE is present, the site preferences would be mainly determined by this factor or OSSE. Thus it is expected that all chromate spinels are normal on account of containing octahedral Cr^{3+}; $MgFe_2O_4$ is essentially inverse, due to the lack of CFSE for Fe^{3+}; $Co_3O_4(\equiv CoO \cdot Co_2O_3)$ is normal since low spin Co^{3+} can gain more CFSE by entering into the octahedral site than Co^{2+} loses locating in tetrahedral site. As for Mn_3O_4 and magnetite Fe_3O_4, the former is normal and the latter is inverse because $Mn^{2+}$and Fe^{3+} has no CFSE in either tetrahedral or octahedral field but Mn^{3+} and Fe^{2+} has a preference for octahedral sites.

3.4.3 Effects of the d Electron Configuration on the Radii of Transition Metal Ions

The reasons for affecting the trend of the radii of octahedrally coordinated divalent transition metal ions shown in Figs. 3.13 and 3.14 are described in the following.

For these three ions of $Ca^{2+}(d^0)$, $Mn^{2+}(d^5, HS)$, and $Zn^{2+}(d^{10})$ with empty, half-filled, and fully-filled orbitals, respectively, the distribution of electron density around the M^{2+} ions looks like spherical symmetry (not corresponding to uniform distribution) by assembling the five d orbitals together (d_{xz} and d_{yz} are not illustrated), shown in the inset of Fig. 3.13. The gradual decrease in radius with increasing atomic number is associated with poor shielding of the nuclear charge due to most d electrons in the t_{2g} orbitals and the compact d orbitals, leading to a greater effective nuclear charge

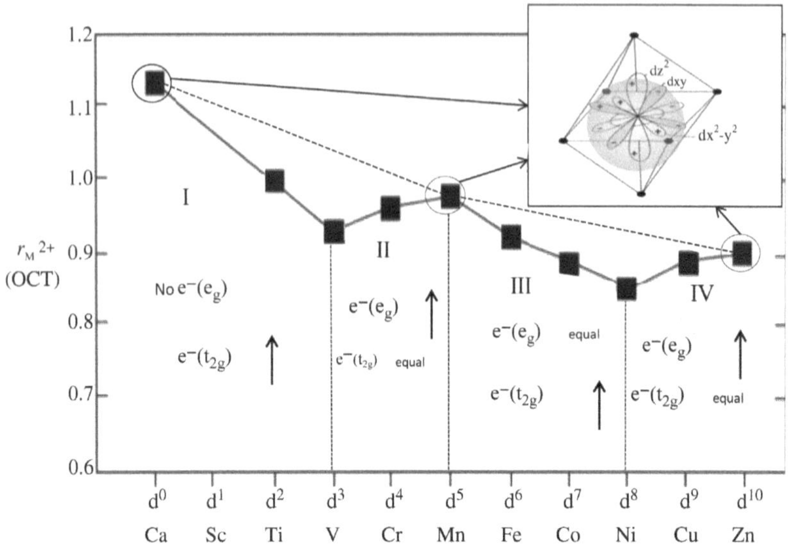

FIGURE 3.13 Radii of octahedral coordinated divalent transition metal ions with high spin states; as the 3d orbitals are empty, half-filled, or fully filled by the electrons, the distribution of electron density around the M^{2+} ions is a spherical-like distribution, shown in the inset but the orbitals of d_{yz} and d_{zy} are not revealed. The 3d transition metals are divided into four regions, and each region indicates the trend of the number of the electron of e_g and t_{2g} orbitals. Arrows indicate the increasing trend.

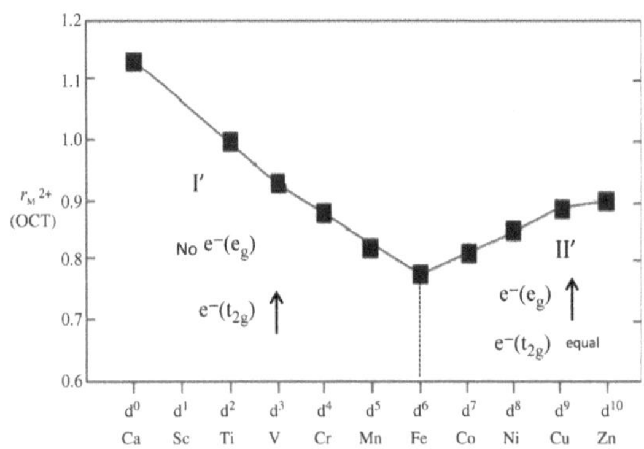

FIGURE 3.14 Radii of octahedral coordinated divalent transition metal ions with low-spin states. The 3d transition metals are divided into two regions, and each region indicates the trend of the number of the electrons of e_g and t_{2g} orbitals. Arrows indicate the increasing trend.

experienced by the outer bonding electrons and resulting in a steady contraction in radius.

The d electron distribution of these transition metal ions with d^1 to d^4 and d^6 to d^9 with low-spin and high-spin states is not spherical. Their radii are smaller than expected, which is attributed to the fact that the shielding of the nuclear charge by these electrons is reduced even further. The variation of the ionic radii of these transition ions would depend on the occupation of their d electrons on t_{2g} or e_g orbitals since t_{2g} orbitals do not shield the bonding electrons, occupying regions of the space that are directed away from the (M^{2+}-anion) axis, but e_g orbitals do shield the nuclear charge much more effectively. The variation of the electrons of e_g and t_{2g} orbitals has been indicated in different regions represented in Figs. 3.13 and 3.14, which makes the trend of the ionic radii with the d electron configuration explicit. It is apparent that, as with the increase of atomic number, the radii increases with the increase of e_g electrons but reduces with the increase of t_{2g} electrons. For example, in comparing V^{2+} to Cr^{2+}(HS) and Cr^{2+}(LS), both Cr^{2+}(HS) and Cr^{2+}(LS) have one extra nuclear charge but the extra electron of Cr^{2+}(HS) is in the e_g orbitals leading to an increased radius and Cr^{2+}(LS) instead in the t_{2g} orbitals giving rise to a decreased radius.

3.4.4 Jahn–Teller Distortions

The Jahn–Teller theorem suggested that in an electronically degenerate state, a nonlinear symmetric molecule could not stay in equilibrium in an orbitals-degenerate state but would rather undergo distortion so that degeneracy would be removed. An electronically degenerate state represents the availability of more than one kind of degenerate orbitals for an electron. In this condition the degenerate orbitals are asymmetrically occupied. On account of the asymmetric occupation, the orbital with more electrons will encounter stronger repulsions, namely, it possesses higher energy than the orbital with fewer electrons. Thus, the ground state of a nonlinear molecule (complex ion) is unstable and has to be nondegenerate or bears a minimum degeneracy implying a distortion of the ionic (molecular) group, so as to bring it to a lower symmetry or energy.

How can one explain why a nonlinear molecule of higher symmetry is not stable? Consider Cr^{3+} ion of $(t_{2g})^3$ with one electron on each orbital, there is just one way of achieving this electron distribution. Thus, the ground state is not degenerate, indicating that the chromium complex may be rigorously octahedral. For Ti^{3+} of a $3d^1$ configuration, the ground state of the electron is of $(t_{2g})^1$ type, which is threefold degenerated since the electron may occupy one of the t_{2g} orbitals in three ways, which is an electronically degenerate state. So $[Ti(H_2O)_6]^{3+}$ is not stable in octahedral symmetry and adopts a lower symmetry.

3.4.4.1 The Jahn–Teller Effect in the Octahedral Field

The Jahn-Teller effect may occur for d^1, d^2, d^4, d^5(LS), d^6(HS), d^7, and d^9 ions in the octahedral field but only d^4(HS), d^7(LS), and d^9 ions undergo pronounced distortions because of the electrons unevenly distributed in

e_g orbitals. Concerning the Jahn-Teller distortion, it is to stabilize the structure in a lower energy or ground state and contemporary with the splitting of degenerate e_g and t_{2g} orbital energy levels of the metal ion. In order to lower the energy, the distance between ligands and e_g orbital lobe of metal ion with more electrons will be lengthened, but shortened for fewer electrons. Therefore, the energy levels of the former will be lowered by 1/2 δ_1 but that of the latter will be raised by 1/2 δ_1 since the electrostatic repulsion between metal ion and ligands decreases with the square of distance. As a result, the system will be stabilized by an amount of $1/2\delta_1$ relative to normal octahedral coordination, as shown in Fig. 3.10, where d^7 (LS) has been set as an example, and assume that d_{z^2} possesses one electron but $d_{x^2-y^2}$ is empty, in which the former is lowered $1/2\delta_1$ but the latter is raised $1/2\delta_1$. Moreover, in accompanying the distortion of e_g orbitals, the energy levels of t_{2g} orbitals will also be changed, where d_{xy} is lifted by $2/3\delta_2$ and d_{xz} and d_{yz} is lowered by $1/3\delta_2$. Thus there is no net change in energy for t_{2g} orbitals. By contrast, if $d_{x^2-y^2}$ possesses one electron but d_{z^2} is empty, the former is lowered $1/2\delta_1$ but the latter is raised $1/2\delta_1$; however, d_{xy} is lowered by $2/3\delta_2$ and d_{xz} and d_{yz} is lifted by $1/3\delta_2$. Thus there is also no net change in energy for t_{2g} orbitals. In addition, δ_2 usually is small, and can be neglected in most systems.

As a result of Jahn-Teller distortion, the metal-ligand bond will be lengthened along the direction of e_g orbital with more electrons, but shortened for fewer electrons. Take d_{z^2} orbital with more electrons, which will lengthen bonds along the *z*-axis (*z*-out distortion); by contrast, if $d_{x^2-y^2}$ orbital with more electrons, the bonds along the *z*-axis will be shortened (*z*-in distortion). While theoretically it is not possible to predict the type of distortion that occurs when the degeneracy occurs in e_g level, it is observed that *z*-out distortion is more preferred. Usually, d^4 (HS), d^7 (LS), and d^9 ions show the z-out distortion.

For Cu^{2+} (d^9), its distorted octahedral has two long and four short bonds when d_{z^2} are doubly occupied but $d_{x^2-y^2}$ singly occupied or four long and two short bonds when d_{z^2} are singly occupied but $d_{x^2-y^2}$ doubly occupied. Thus the distortion of the related Cu compounds is varied. For example, CuF_2 with a distorted rutile structure wherein four fluorine ions at 1.93 Å and two at 2.27 Å, $CuCl_2$ with four chlorine ions at 2.30 Å and two at 2.95 Å, and CuO with an effectively square planar structure wherein four oxygen ions at 1.95 Å and two at 2.88 Å.

3.4.4.2 Development of Square Planar Coordination

Concerning the d^8 configuration, it usually is misunderstood that there is no distinction between the low spin and the high spin because of the same configurations. However, as observed in Fig. 3.15, Δ_o basically is different: low spin is high but high spin is low. Moreover, the square planar structure is often observed in low spin with high Δ_o, which occurs frequently in $3d^8$ ions with strong field ligands and $4d^8$ and $5d^8$ ions with any ligands, due to the fact that their orbitals are large and more diffuse and extend to greater

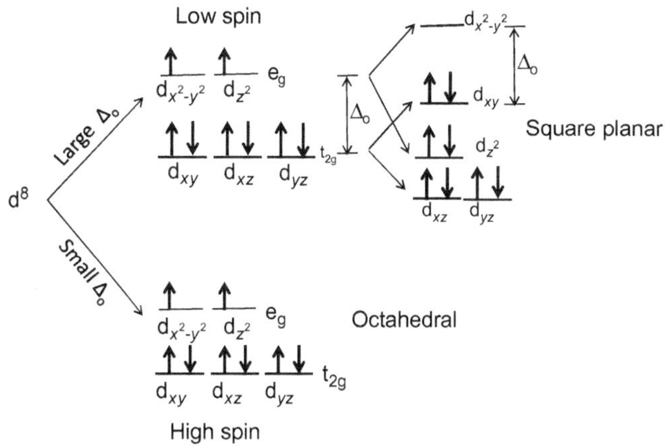

FIGURE 3.15 For transition metals with d^8, typically, the two e_g electrons of d_{z^2} and $d_{x^2-y^2}$ orbitals are singly occupied even though the values of Δ_o are different, low for high spin but high for low spin. As for low spin, on account of the energy consideration, the two e_g electrons of d_{z^2} or $d_{x^2-y^2}$ orbitals may become doubly occupied, leading to the square planar coordination, where the energy level of d_{z^2} is lower than that of d_{xy}.

radial distances from the nucleus, leading to the better overlap between the metal lobes and ligands. It is inferred that high repulsion force between the metal lobes and ligands would be the main reason for developing the square planar coordination. As the square planar structure develops, diamagnetism is also observed, which is attributed to the d_{z^2} orbitals being doubly occupied. However, the pairing of the electrons occurring in the e_g orbitals actually violates Hund's rule. Nevertheless, the energy benefit may promote such configuration. As the d_{z^2} orbital is doubly occupied, it encounters higher repulsion force. In order to stabilize the system, it is lengthened, and in turn its energy level is lowered, stabilized by an amount of $1/2\delta$ relative to normal octahedral coordination. In such a circumstance, competition between δ and the pairing energy, P, shall be taken into account. For small elongation along z-axis, i.e., δ is small, P may be higher than δ so that there would be no gain in stability for the d_{z^2} orbital to be doubly occupied. Therefore, large δ becomes necessary, requiring the very high repulsion force between the metal lobes and ligands. The doubly occupied d_{z^2} orbital will become stable until reaching where $P < \delta$, and even resulting in an infinite distortion of octahedral structure along the z-axis, equivalent to removing the z-axis ligands and leading to square planar coordination. As shown in Fig. 3.15, the energy level of the d_{z^2} orbital is even lower than that of the d_{xy} orbital. In addition, the lowest four orbitals are filled and the highest orbital of $d_{x^2-y^2}$ is empty, resulting in a large CFSE. In regard to square planar structure, the ligands are absent along the z-axis, and focused on the xy plane, the energy difference between the energy levels of $d_{x^2-y^2}$ and d_{xy} orbitals is equal to Δ_o. It is noticed that this, however, is not the Jahn–Teller distortion because the octahedral d^8 is not electronically degenerate, and in fact, there are many cases without axial ligands along the z-axis. In the case of high spin, i.e., Δ_o is not very high, and the two e_g electrons singly occupying d_{z^2} and $d_{x^2-y^2}$ orbitals, respectively, which are degenerate, the coordination remains normal octahedral. Because of the unpaired electrons, the compounds are paramagnetic, shown at the bottom of Fig. 3.15.

For ionic oxide crystals, since δ increase in the order of 3d, 4d, and 5d, the Δ_o value increases by about 50% down the group. The rocksalt structure of NiO (3d ion) possesses normal octahedral but PdO (4d) and PtO (5d) possess square planar coordination. However, PdF_2 with the rutile structure prefers in octahedral coordination, because its ligand (F^-) results in small δ.

3.4.4.3 *The Jahn–Teller Effect in the Tetrahedral Field*

In the tetrahedral field, Jahn–Teller distortion occur when the upper t_2 orbital contain 1, 2, 4, or 5 electrons, i.e., d^3 (HS), d^4 (HS), d^5 (LS), d^8, d^9. Assuming that there is one electron or four electrons in the t_2 orbitals and taking the z-axis as the tetrahedral axis, either electron distribution can be considered to be equivalent to one electron present in the d_{xy} orbital but absent in others, which will repulse ligands along the z-axis. Thus in order to stabilize the structure in a lower energy or ground state, Jahn-Teller effect will cause the stretch of the tetrahedral coordination, as shown in Fig. 3.16A. If there are two or five electrons in the t_2 orbitals, to preserve tetrahedral symmetry, the two electrons would lie singly in the d_{xz} and d_{yz} orbitals, respectively, but the five electrons would occupy these three orbitals, e.g., one in d_{xy} and double in both d_{xz} and d_{yz}. This implies that both electron distributions are equivalent to a positive hole present in the d_{xy} orbital but absent in others, which will attract ligands along the z-axis. Thus for the sake of stabilizing the structure in a lower energy or ground state, Jahn-Teller effect will result in the flattening of the tetrahedron, as revealed in Fig. 3.16B.

Here, two examples are given to show how Jahn-Teller effect influences the structures of spinels, which usually have cubic symmetry:

1. $CuFe_2O_4$ with an inverse spinel structure, where c/a = 1.06: Because Cu ions located in the octahedral sites would lead to two Cu–O bonds along

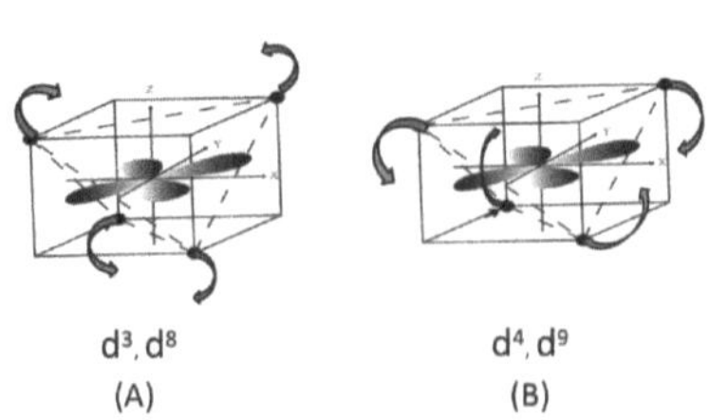

FIGURE 3.16 Structure distortion in a tetrahedral field with (A) one electron or four electrons in the t_2 orbitals, leading to stretched tetrahedral coordination and (B) two or five electrons in the t_2 orbitals, leading to flattened tetrahedral coordination.

z due to the Jahn–Teller effect, the bond length along z-axis would be longer than four Cu–O bonds in the xy plane.

2. $CuCr_2O_4$ with a normal spinel structure, where $c/a = 0.91$: In this case, Cu ions located in tetrahedral sites whose bonds are flattened in the z direction due to the Jahn–Teller distortion would lead to a shorter c-axis.

3.4.5 Structure Distortions Arising From Asymmetric Electron Density or Inert Pair Effect

The structural distortion may arise from the inert (lone) pair effect, referring to the reluctance of the heavy, posttransition elements, especially Group 13 (Tl), Group 14 (Pb, Sn), and Group 15 (Bi), with the asymmetric electron densities. The lone pair of $6s^2$ electrons is usually considered to be stereochemically active in that these electrons are not in a spherically symmetrical orbital but stick out to one side of the cation (perhaps in some kind of s-p hybridized orbital). Different kinds of distortions may occur:

1. The lone pair may locate between the cation and its nearest anionic neighbors, causing a distorted polyhedral with the variation of bond length of the cation. However, a report (Walsh and Watson, 2005) suggested that this explanation may not be valid in comparing PbO with PbS, where the former has distortion but the latter does not. They proposed that the distortion actually arose from the net asymmetry in the electron density on Pb due to the coupling of Pb 6p with the antibonding Pb 6s–anion p states rather than the lone pair effect per se. Because energy states of the Pb 6s and 6p are too far separated to couple directly, this coupling can only occur as an appropriate anion could interact with Pb 6s to make Pb 6s states close to the Fermi level. Since oxygen anions can meet the related requirement but sulfur anions cannot, PbO would reveal distortion but PbS would not.
2. The lone pair may take the place of the plane of an anion and its associated pair of bonding electrons. Since a nonbonding pair of electrons occupies a larger volume of space than a bonding pair, it would cause distortions of the polyhedral.

3.5 STRUCTURE OF MATERIAL TECHNOLOGICAL INTEREST

This section actually is dealing with the ceramic structures. The majority of ceramic structures can be characterized by either face-centered-cubic or hexagonal close packing of anions with the fractional occupancy of each of the interstitial sites that are defined by that anion packing. In the following, the ceramic structures defined on the basis of their stoichiometry will be briefly described in the order of the coordination number (CN) of cations, listed in Table 3.2.

TABLE 3.2 Ionic and Covalent Structures Categorized According to the Stoichiometry, Anion Packing, and Coordination Number

Stoichiometry	Structure	Anion Packing	Coordination Number of Cations	Materials
AX	Zinc blende	FCC	4	β-ZnS, AgI, β-SiC
	Wurtzite	HCP	4	α-ZnS, α-SiC, BeO, ZnO, AlN
	Rocksalt	FCC	6	NaCl, MgO, MnO, FeO, CoO, CaO, NiO
	Nickel arsenide	HCP	6	NiAs, FeS, CoSe
	Cesium chloride	Simple cubic	8	CsCl, CrBr, CsI, TiCl
AX_2	Silicates	Corner-shared SiO_4 tetrahedra	4	SiO_2
	Cadmium iodide	HCP	6	TiS_2, MgI_2
	Rutile	Distorted HCP	6	TiO_2, SnO_2
	Fluorite	Simple cubic	8	CaF_2, SrF_2, ThO_2, ZrO_2
	Antifluorite	FCC	4	Li_2O, Na_2O, K_2O
A_2X_3	Corundum	HCP	6	Al_2O_3, α-Fe_2O_3, Cr_2O_3, Ti_2O_3, V_2O_3
AB_2X_4	Spinel	FCC	4, 6	$MgAl_2O_4$
$B(AB)X_4$	Inverse spinel	FCC	4, 6	$FeMgFeO_4$
ABX_3	Ilmenite	HCP	6, 6	$MgTiO_3$, $FeTiO_3$
	Perovskite	FCC (AO sublattice)	12, 6	$CaTiO_3$, $BaTiO_3$

The structures of the important materials of technological interest with diverse applications will be illustrated and discussed in the following.

3.5.1 Structure and the Related Feature of ZnO

ZnO is a II–VI compound semiconductor whose ionicity resides at the borderline between the covalent and ionic semiconductors. In ZnO, the ionic radius of Zn^{2+} is large enough to be coordinated with 6 O^{2-}, but because of sp^3 in the Zn–O bond, Zn^{2+} occupies the tetrahedral site. According to the first-principles calculations, the strong Coulomb repulsion between Zn and O, due to the 3s, 3p, and 3d (d^{10}) states of the Zn ion, favorably causes the shift of the Zn ion to the O ion. Thus, different crystal structures of ZnO can be synthesized as per processing conditions: e.g., wurtzite, zinc blende, and rocksalt, as schematically represented in Fig. 3.17.

Under ambient conditions, the thermodynamically stable phase is of wurtzite structure. The zinc blende ZnO structure can be stabilized only by growth on cubic substrates, and the rocksalt structure may be developed at relatively high pressures.

For material science applications, zinc oxide (ZnO) has a high refractive index, high thermal conductivity, binding, antibacterial, and UV-protection properties. Thus ZnO as a vital semiconductor with tremendous scientific and technological interest has an outsized exciton-binding energy (60 meV) and wide direct band gap (3.37 eV or 375 nm at room temperature), and it is

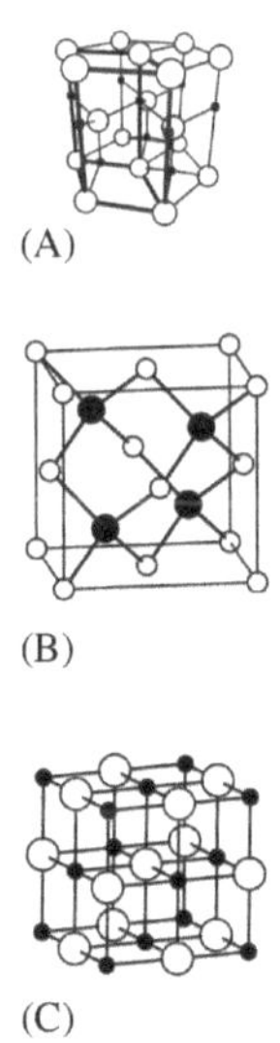

FIGURE 3.17 Structures of ZnO with different forms: (A) wurtzite, (B) zinc blende, and (C) rocksalt. Solid spheres: Zn; empty spheres: O.

considered to be a possible material in optoelectronic applications to operate in the visible and close to ultraviolet spectral regions. Therefore, the most common potential applications are in laser diodes and light-emitting diodes (LEDs); and it also is widely utilized in several industrial areas such as UV light-emitting devices, ethanol gas sensors, photocatalysts, pharmaceutical, and cosmetic industries. Aluminum-doped ZnO layers are used as a transparent electrode. One application which has begun to be commercially available is the use of ZnO as the front contact for solar cells or of liquid crystal displays. Transparent thin-film transistors can also be produced with ZnO. Currently, one of the main applications of nanoparticle ZnO is as an effective drug delivery system.

3.5.2 Structure and the Related Feature of SiC

Silicon carbide presents as about 250 crystalline forms, whose polymorphism is characterized by a large family of similar crystalline structures called polytypes, which can be viewed as layers stacked in a certain sequence, shown in Fig. 3.18.

The most common polymorph is alpha silicon carbide (α-SiC) which has a hexagonal crystal structure similar to Wurtzite, whereas β-SiC has a zinc blende crystal structure. Its use was considered in the past to be in structural applications but currently much interest is in its use as a semiconductor material in electronics by doping acceptors or donors due to its high thermal conductivity, high electric field breakdown strength, high maximum current density, and a very low coefficient of thermal expansion (4.0×10^{-6}/K). SiC

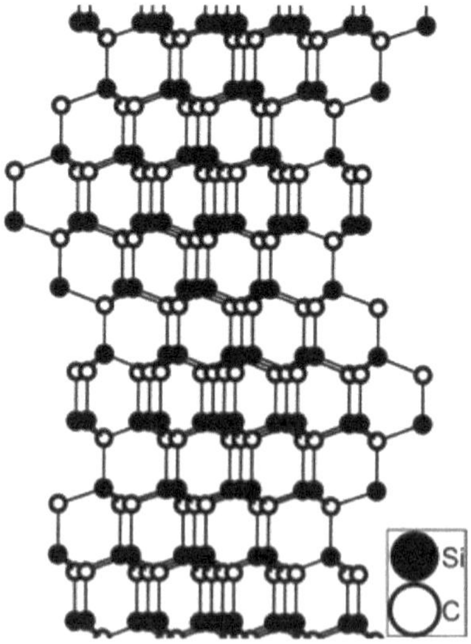

FIGURE 3.18 Structure of SiC.

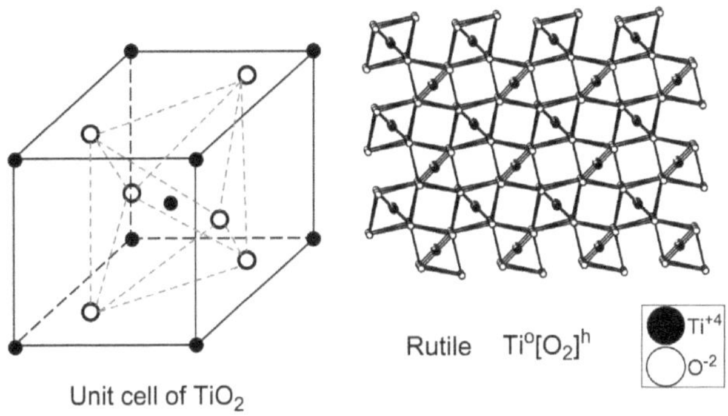

FIGURE 3.19 Rutile structure of TiO_2.

has been used as varistor, ultraviolet LED, Schottky diodes, and high-power switching. It has been considered as one of the promising methods to synthesize graphene on a large scale for practical applications.

3.5.3 Structure and the Related Feature of TiO_2

Among the polymorphs, TiO_2 with its rutile structure is the most applicable, in which only half the octahedral sites are filled by cations in a distorted HCP lattice, shown in Fig. 3.19.

Such a structure essentially makes cation interstitials become possible and this is evidenced in its application as an oxygen sensor. Titanium dioxide is the most commonly studied photocatalyst due to its chemical stability, eco-friendliness, nontoxicity, and cost-effective synthesis. Other applications include solar cells and batteries.

3.5.4 Structure and the Related Feature of ZrO_2

Three phases of ZrO_2 are known: monoclinic $<1170°C$, tetragonal 1170–2370°C, and cubic $>2370°C$. The trend is for higher symmetry at higher temperatures, as is usually the case. Unlike TiO_2, which features six-coordinate Ti in all phases, monoclinic zirconia consists of seven-coordinate zirconium centers. This difference is attributed to the larger size of the Zr atom relative to the Ti atom. A few percentage of the oxides of calcium or yttrium stabilize the cubic phase with a fluorite structure shown in Fig. 3.20, in which the octahedrons are distorted due to the ionic size which is relatively smaller than the space of the octahedral sites.

Thus, stabilized zirconia not only produces a high oxygen vacancy concentration but also possesses an open distorted structure with a small energy of migration (~ 1 eV) resulting in an unusually high oxygen diffusion coefficient and allowing oxygen ions to move freely through the crystal structure at high temperatures. However, for highly doped or highly nonstoichiometric

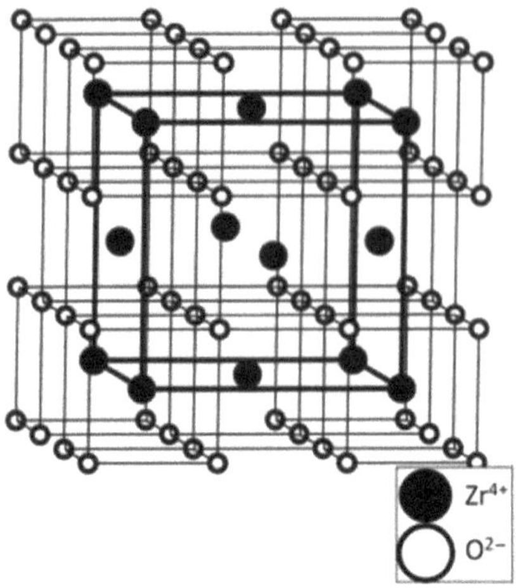

FIGURE 3.20 Fluorite structure of ZrO_2.

systems, simple association, formation of compact clusters containing Ca_{Zr}'' and $V_o^{\cdot\cdot}$, the ordering of defects into extended defects, e.g., crystallographic shear phases, and long-range ordering of oxygen vacancies may decrease the concentration of free oxygen vacancies and in turn cause the oxygen ion conductivity to decrease remarkably. Nevertheless, the high ionic conductivity and a low electronic conductivity makes it useful in oxygen sensor and fuel cell membranes, and it is also used as the solid electrolyte in electrochromic devices. Moreover, its very low thermal conductivity has led to its use as a thermal barrier coating in jet and diesel engines to allow operation at higher temperatures.

3.5.5 Structure and the Related Feature of Spinel (AB_2O_4)

The Bravais lattice of the conventional unit cell of the spinel structure is FCC; there are eight formula units per cubic unit cell, consisting of 32 anions and 24 cations, for a total of 56 atoms. The anion sublattice of most spinels is arranged in a pseudo-CCP spatial arrangement. There are 96 interstices between the anions in the cubic unit cell; however, in AB_2O_4 compounds, only 24 are occupied by cations. Of the 64 tetrahedral interstices that exist between the anions, 8 are occupied by cations. The remaining 16 cations occupy half of the 32 octahedral interstices.

The eight formula units can further be classified into two types, shown in Fig. 3.21: In subcubes α, the A cations occupy half of the eight corners diagonally as well as at the center of the subcube. The cations at the center of the subcube are tetrahedrally coordinated, where the four anions occupy half of the eight corners diagonally.

In subcubes β, the A cations occupy half of the eight corners of the subcube diagonally. The subcube contains four B cations octahedrally

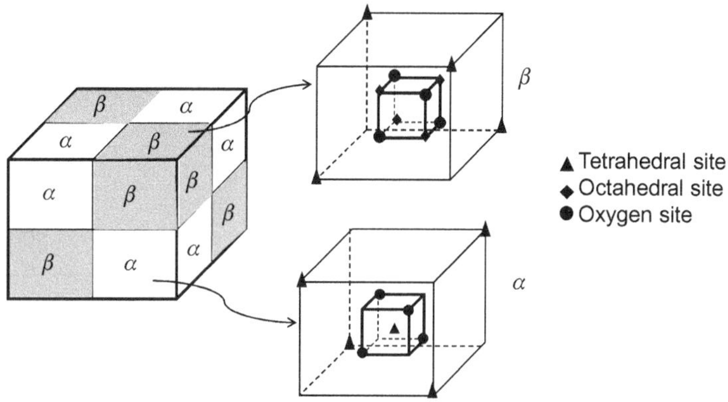

FIGURE 3.21 Spinel structure.

coordinated with anions and located at half of the eight corners diagonally, and four anions located at the remaining four corners.

The formula in terms of 32 anions in a unit cell of the spinel structure can be written as $(Tetra.)_8[Oct.]_{16}X_{32}$, corresponding to the usual chemical formula of AB_2X_4. In a "normal" spinel, the tetrahedral sites are occupied by the A atoms, and in an "inverse" spinel they contain (half of the) B ions. Octahedral sites in a normal spinel are the B ions, whereas in an inverse spinel, these sites contain the A ions and half of the B ions, e.g., normal spinel ($ZnFe_2O_4$): $(A)[B]_2X_4$ and inverse spinel ($NiFe_2O_4$): $(B)[AB]X_4$.

3.5.5.1 $LiMn_2O_4$

The stoichiometric $LiMn_2O_4$ with a normal spinel structure shows a reversible phase transition from cubic to orthorhombic structure around room temperature of about 290 K (Fang and Chung 2011). Such a transformation is attributed to Jahn–Teller distortion of high spin Mn^{3+} ions with accompanying columnar charge ordering in the manganese sublattice, so-called electronic crystallization (Rodriguez-Carvajal et al., 1998). Upon cycling, such behavior would cause the decrease of the cell capacity. While it is possible to suppress the phase transition by partial substitution of Mn^{3+} by 3d transition metal ions or Al^{3+}, the insertion of lithium excess in octahedral 16d sites in the spinel structure can stabilize the structure. The spinel-type oxide $Li_{1+x}Mn_{2-x}O_{4-\delta}$ and its related materials have been widely used as cathode materials for secondary lithium ion batteries.

3.5.5.2 Magnetite

Magnetite (Fe_3O_4) undergoes phase transformation accompanied by a complex structural distortion and becomes electrically insulating at temperatures less than 125 K, shown in Fig. 3.22.

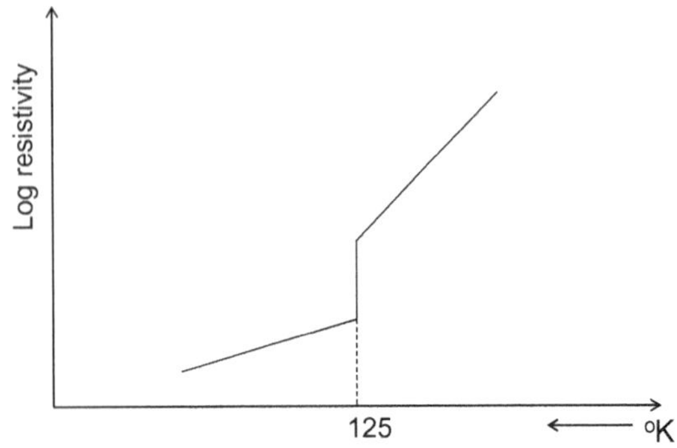

FIGURE 3.22 Schematic illustration of the electrical resistivity as a function of temperature for a nearly stoichiometric magnetite, showing the Verwey transition temperature at about 125 K.

The well-known Verwey transition in magnetite has been under research for a century. At the Verwey transition temperature ($T_V \sim 125$ K), magnetite transforms from cubic ($T > T_V$) to monoclinic ($T < T_V$) symmetry. This transformation is related to changes in crystal symmetry and in cation ordering of Fe^{2+} and Fe^{3+} ions along with dramatic property changes like in electrical conductivity and heat capacity.

Each B-site FeO_6 octahedron has an irregular distortion arising from the bimodal distribution of tetragonal distortions and their correlation with the radial amplitudes due to charge ordering and associated orbital-ordering displacements. The inverse-spinel arrangement accounts for the intensity of spontaneous magnetization in magnetite since the B sites contain twice as many Fe atoms as the A sites, leading to a fairly strong net spontaneous magnetization. Moreover, the inverse spinel structure contributes to magnetite's weak magnetocrystalline anisotropy at $T > T_V$. Magnetocrystalline anisotropy in general arises from a combination of single-ion anisotropy and anisotropic exchange interactions, related to the way that electron orbitals fit into the crystal lattice. The inverse spinel cation arrangement (effectively lacking distinct ferrous ions due to B-site electron hopping), together with the cubic crystal symmetry, accounts for the relatively weak magnetocrystalline anisotropy of magnetite at room temperature.

While the transition was suggested (Jackson et al., 2011) fundamentally as a coordination crossover—a reversible transformation between normal ($T < T_V$) and inverse ($T > T_V$) spinel arrangements—other models generally involve some form of incomplete charge ordering of the octahedral iron ions, in combination with ordering of electron-orbital orientations of the ferrous ions, lattice distortion, and other additional complicating factors, and the arguments revolve around the relative importance of the various interacting effects. However, a recent report (Senn et al., 2012) suggested that the localized electrons distributed over linear three-Fe-site units, called "trimerons" may cause anomalous shortening of some Fe–Fe distances. The charge order and three-site distortions induces substantial off-center atomic displacements resulting in the large electrical polarization. The coupling of electrical polarization to the magnetization may play a significant role in the Verwey transition. The current significant application is the use of nanoparticles of magnetite in drug delivery systems.

3.5.6 Structure and the Related Feature of Perovskite ABO_3

The structure of perovskite, e.g., $CaTiO_3$ or $BaTiO_3$, can be described in two different ways, illustrated in Fig. 3.23: an A cation is located at the center of the cube coordinated with 12O anions (black lattice) and a B cation is placed at the center of the cube octahedrally coordinated with 6O anions (red lattice).

The geometrical configuration for the arrangement of the octahedral polyhedral with the lowest interaction energy is the corner-sharing connection. Nevertheless, most perovskites are distorted and do not have the ideal cubic structure. Three main factors are identified as being responsible for the distortion: size effects, the effect of defect (anion deficiency), and the Jahn–Teller effect.

a. Size Effects

In an ideal perovskite crystal, ionic radii of A, B, and O satisfy the following relationship

$$r_A + r_O = (2)^{1/2}(r_B + r_O)$$

For real case,

$$r_A + r_O = t(2)^{1/2}(r_B + r_O),$$

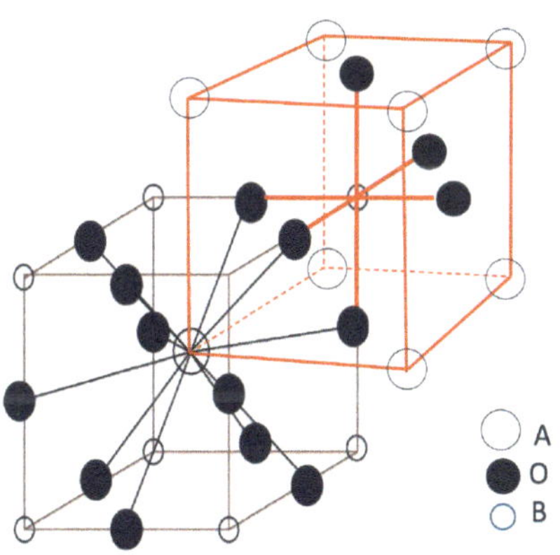

FIGURE 3.23 The structure of perovskite. Black lattice: an A cation is located at the center of the cube coordinated with 12O anions. Red lattice: a B cation is placed at the center of the cube octahedrally coordinated with 6O anions.

where t is the tolerance factor. Since the bonding of perovskites is not truly ionic and also because of the dependence on what value of the ionic radii is used in the equation, the tolerance factor, t, is actually a rough estimate. In general, the tolerance factor of the perovskite structure lies within the range $0.75 < t < 1.00$. In the case where t is larger than 1 owing to a large A or a small B ion, a hexagonal-type perovskite structure would occur, e.g., $BaNiO_3$, with t value 1.13 ($r_A = 1.61$ Å and $r_B = 0.48$ Å) and face-sharing of the $[NiO_6]$ octahedra.

Moreover, it should be noted that the value of t may also vary with temperature due to the phase transition. A wide variety of structural instabilities may display in the various materials. These may involve rotations and distortions of the O octahedral as well as displacements of the cations from their ideal sites. The interplay of these instabilities accounts for the rich variety of ferroelectricity (FE) and antiferroelectricity (AFE) behaviors. The two common instabilities result from softening of either a polar zone-center phonon mode, leading to a FE phase, or the softening of a nonpolar zone-boundary mode involving rotations of oxygen octahedra, leading to an antiferroelectro-distortion phase. In some cases, a zone-boundary polar mode may also occur, leading to an AFE phase. Regarding ferroelectric ABO_3, the most stable structure is tetragonal for $t \geqq 1$, cubic for $t \approx 1$, and rhombohedral or orthorhombic for $t \leqq 1$, e.g., both $BaTiO_3$ with $t = 1.062$ and $PbTiO_3$ with $t = 1.019$ show tetragonal structures at room temperature. Since FE is strongly correlated with the B-site cations, it was usually considered that the size of A-site cations might influence the ferroelectric properties, e.g., the variation of the Curie temperatures between $BaTiO_3$ and $SrTiO_3$ has usually been attributed to their varying the space of B sites; however, this viewpoint is inappropriate. While the origin of the ferroelectricity is still not completely clarified (Benedek, 2014), the first principle calculation demonstrated that FE in the perovskites resulted from the fact that the long-range forces due to the dipole–dipole interaction which favored the FE state overcame the short-range forces due to the Coulomb repulsions which favored the cubic phase. It has further been proposed (Cohen, 1992) that the hybridization between Ti 3d state and O 2p state or instead by recent report (Miura et al., 2011)between Ti 3s and 3p states and O 2p states in $BaTiO_3$ and $PbTiO_3$, which weakens the short-range force of the Coulomb repulsions between Ti and O ions, may play a crucial role in developing FE. Actually, $CaTiO_3$ also possesses the hybridization between Ti 3d and O 2p, which however, is not strong enough for it to reveal FE. Since its dielectric constant increases with decreasing temperature and saturates at low temperatures, it has been classified as an incipient ferroelectric or a quantum paraelectric material similar to $SrTiO_3$, $KTaO_3$, and TiO_2. Ferroelectric materials with high dielectric constants well known as electroceramics are the significant application in the capacitor industry, especially in multilayer ceramic capacitors.

b. The Effect of Defect (Anion Deficiency)

The oxides with perovskite structures very commonly become anion-deficient during processing at high temperatures. The anion deficiency may vary the size and the type of the unit cell and the coordination number of the B sites, which leads to modification of the band structures and properties. The chemistry of the anion-deficient perovskites can be represented by the general formula $A_{(m+n)-yn}B_{(m+n)}O_{3(m+n)-xn}$ with $0 \le x < 1$, $0 \le y < 1$, $0 \le n < m$, and n and m integers. For example, $SrFeO_{\delta}$ with δ varying between 2.5 and 3, take $\delta = 2.875$, i.e., $SrFeO_{2.875}$, in which the oxidation states of Fe ions can be assigned to +3 and +4. Moreover, because of the ordering of oxygen vacancies, FeO_5 square pyramids are formed. Several other types of vacancy orderings are also known according to the homologous series $A_mB_mO_{3m-1}$, e.g., $Ca_2Mn_2O_5$ and $La_2Ni_2O_5$ with $m = 2$.

c. Jahn–Teller Effects

If the cations in the B sites of the perovskite structures are Jahn–Teller active transition metals described in Section 3.4.4, the distortion of the structure can be attributed to Jahn–Teller effects, e.g., in $LnMnO_3$ (Ln = La, Pr or Nb) with Mn^{3+} ions having four electrons in the 3d orbitals.

Since most of the newly discovered materials with so-called colossal properties are relevant to the perovskite or perovskite-related structures, we will illustrate them in the following.

3.5.6.1 $BiFeO_3$

There are about eight structural transitions and a weak F ordering in its ground state structure having been suggested in $BiFeO_3$ (Catalan and Scott, 2009). As aforementioned, the structural instability for perovskite-like oxides can be assessed in terms of the empirical Goldsmith tolerance factor t. Using the ionic radii given by Shannon and Prewitt, with Bi^{3+} in eightfold coordination (the value for 12-fold coordination is not reported) and Fe^{3+} in sixfold coordination and high spin, the calculated value of t is 0.88 for $BiFeO_3$. Since the value of t is smaller than 1, the Bi ion may be unstable in the high-symmetric location possessed by the cubic perovskite-like structure. In order to minimize the lattice mismatch and gain stability, the oxygen octahedra must buckle in order to fit into a cell that is too small, whose rotation angle is about 11–14° around the polar [111] axis, with the directly related Fe–O–Fe angle, about 154–156°. Because of the oxygen shifts in the ideal perovskite (111) plane, Bi ions shift 0.62 Å normal to this plane, leading to the reduction of three Bi – O bond lengths and the increase of the other three that results in an octahedral distortion with rhombohedral symmetry. Thus the structure of the room temperature phase of $BiFeO_3$ is identified as rhombohedral with point group **R3c**, whose lattice parameter and rhombohedral angle of the perovskite-like unit cell are 3.965 Å and ca. 89.3–89.4° at room temperature, respectively, shown in Fig. 3.24. The unit cell can also be described in a hexagonal frame of reference, with the hexagonal *c*-axis [001] parallel to the diagonals of the perovskite cube [111]. The lattice parameters of the hexagonal cell are 5.58 Å and 13.90 Å for a and *c*-axis, respectively.

The Fe–O–Fe angle is significant because it predominates both the magnetic exchange and orbital overlap between Fe and O, which in turn determines the magnetic ordering temperature and the conductivity. The spontaneous polarization is along [111] pseudocubic. The giant polarization is attributed to both the large Fe-ion displacement relative to the negative charge center and the Bi 6s lone pair electrons. Such a result suggests that a large B-site-driven ferroelectric polarization can be induced by the epitaxial strain in BFO (Zhang, 2011).

Because $BiFeO_3$ possesses very large remnant polarization, 100 $\mu C\ cm^{-2}$ along the polar [111] direction, it has been expected to be used in the applications of FE and piezoelectricity, and tetrahertz radiation. However, the most important applied research on $BiFeO_3$ are magnetoelectric and spintronic applications, in which memories are written using a voltage and read using a magnetic field. The multiferroic and optoelectronic properties of $BiFeO_3$ make it a potential candidate for next-generation ferroelectric random access memories, magnetic sensors, and photovoltaics.

3.5.6.2 Sr_2FeMoO_6

If all three unit cell edges of the cubic perovskite structure are doubled, it is possible to occupy equivalent positions with atoms of different elements, e.g., Sr_2FeMoO_6 (SFMO), illustrated in Fig. 3.25.

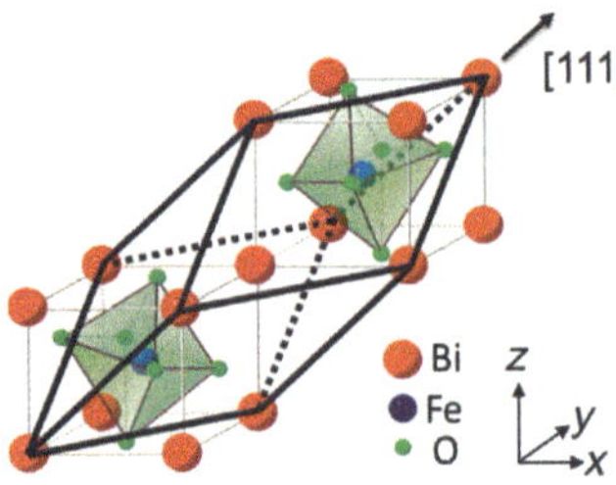

FIGURE 3.24 Rhombohedral structure of $BiFeO_3$.

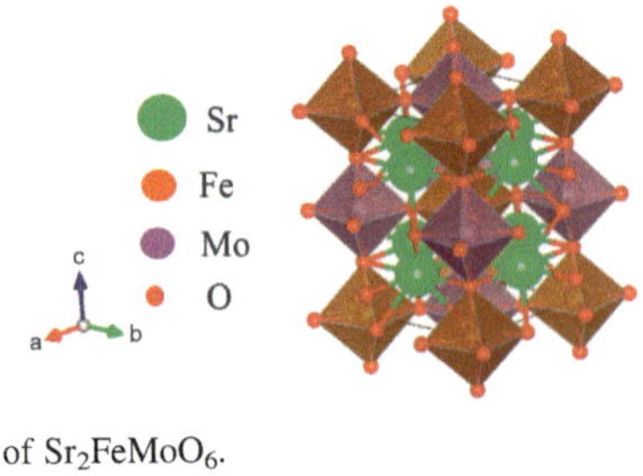

FIGURE 3.25 Structure of Sr_2FeMoO_6.

Corner-sharing octahedra of FeO_6 and MoO_6 construct three-dimensional networks in the structure of SFMO, and the Sr cations locate between the octahedral sites. The ordering of Fe and Mo can be capable of influencing various properties of SFMO drastically. Disorder alone can reduce magnetic ordering as well as control transport properties, giving rise to a metal–insulator transition at a critical level of disorder. Especially, the simultaneous presence of disorder and strong electron–electron correlation effects would lead to many counterintuitive phenomena and exciting physics.

The most striking structure feature is the development of antiphase boundaries (APBs). It has been found that APBs play a significant role in magnetic properties of SFMO, in which ferromagnetic long-range ordered domains coupled antiferromagnetically across APBs may lower remanence in magnetization. Two types of APBs, high-density maze-type in the core and low-density swirl-type in the shell of grains, were for the first time discovered to be simultaneously existent by the author's group (Yang and Fang, 2012). The former is linked together but the latter reveals separate ribbons, whose atom arrangements are confirmed to be zigzag; especially, the antiferromagnetic configuration of Fe ions can be assumed to be exactly located at this type of APB. At a significant fraction of APBs, they have a strong influence on the spin-polarized conduction electrons of SFMO due to the presence of an antiferromagnetic coupling.

The structure of APBs of a sample heated to 1350°C in a mixing gas of 1% H_2/99% N_2 for 4 h was scrutinized by high-resolution transmission electronic microscope (HRTEM). The results of microstructure observation are shown in Fig. 3.26.

Fig. 3.26A reveals a core–shell structure in a grain, wherein the shell part near grain boundaries contained only a few APBs with a swirl-type; by contrast, there were a huge amount of maze-like APBs with nanosized domains existing in the core of a grain. Moreover, the morphologies of these two APBs are different: the maze-type APBs are linked together but swirl-type APBs reveal separate ribbons. This figure is a center dark field image recorded along [010] zone axis, and the electron beam was tilted until the (101) spot was centered. Fig. 3.26B reveals no significant observation of APBs but a few stacking fault (α-boundaries) existing inside the grain. This figure is a bright-field image recorded along [010] zone axis. The diffraction pattern shown in Fig. 3.26C confirms that the images in (A) and (B) indeed were recorded along [010] zone axis. It should be stressed that high contrast of APBs cannot be clearly observed by bright-field image (Fig. 3.26B), and the center dark field image should be employed, confirmed by Fig. 3.26A.

In order to gain a deep insight of atom arrangement of APBs, the detailed analyses of the structure of APBs were further undertaken, shown in Fig. 3.27.

Fig. 3.27A is the center dark field image of a magnified swirl-type APB, and Fig. 3.27B is the high-resolution lattice image of a swirl-type APB

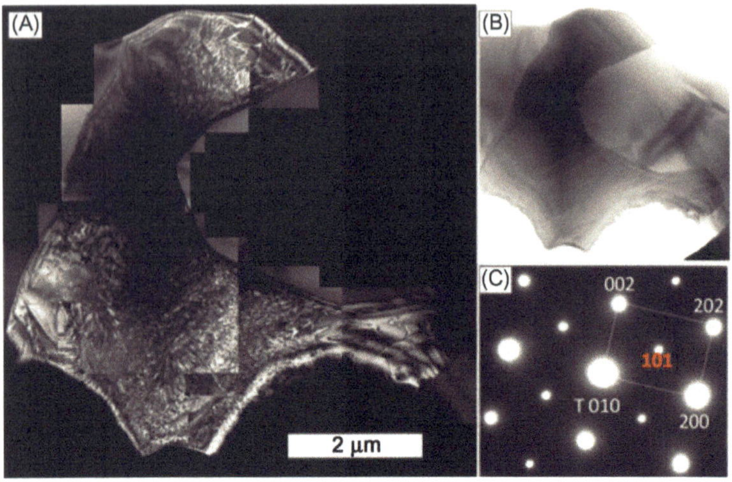

FIGURE 3.26 HRTEM observations of APBs in a core–shell grain of a Sr_2FeMoO_6 sample heated at 1350°C in an atmosphere of 1% H_2/99% N_2 for 4 h. (A) The center dark field image was recorded along [010] zone axis, and the beam was tilted until the (101) spot was centered, showing a large amount of APBs. (B) The bright-field image was recorded along [010] zone axis, revealing no significant observation of APBs. (C) The diffraction pattern confirmed that the images in (A) and (B) indeed were recorded along [010] zone axis. *From Yang, C.W., Fang, T.T., Structures and Development Mechanism of the Anti-Phase Boundaries in Sr_2FeMoO_6 2012. J. Electrochem. Soc. 159 (3), 35.*

marked by a red rectangle in Fig. 3.27A. Fig. 3.27C is the diffraction pattern derived from the FFT of the lattice image in Fig. 3.27B where the supper-lattice spots have been marked. The B-site lattice image was further specified in Fig. 3.28, which can be derived from the inverse FFT image by masking the supper-lattice spots of (**101**), ($\mathbf{10\bar{1}}$), ($\mathbf{\bar{1}01}$), and ($\mathbf{\bar{1}0\bar{1}}$) indicated in the FFT image (Fig. 3.27C).

The derived B-site lattice image (Fig. 3.28A) was magnified in order to clarify the atom arrangement of B sites represented by white balls (Fig. 3.28B) whose d-spacing is 0.46 nm consistent with that of {**101**} planes. By overlaying an ideal lattice of {**101**} plane of SFMO with B sites represented by red and green balls, lattice-mismatch sites marked by the red balls encircled by yellow color can be identified along two orthogonal planes of (**101**) and ($\mathbf{10\bar{1}}$) (or ($\mathbf{\bar{1}0\bar{1}}$) and ($\mathbf{\bar{1}01}$)), wherein a zigzag APB is clearly defined. The established zigzag APB is fully confirmed, in which only two of four superlattice spots were selected in FFT image. While the ions (Fe or Mo) constructing APBs cannot be determined, it was suggested that Fe–O–Fe interactions across APBs with strong antiferromagnetic coupling would be the most possible atom arrangement. In Fig. 3.28C, we show that the antiferromagnetic configuration of Fe ions can be exactly located at zigzag APB, defined in Fig. 3.28B. The zigzag APBs could also be established

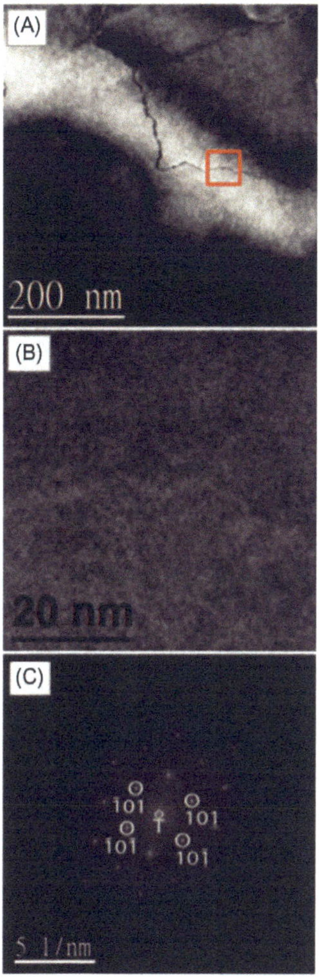

FIGURE 3.27 HRTEM observations of microscopic structure of APBs of a Sr_2FeMoO_6 sample heated at 1350°C in an atmosphere of 1% H_2/99% N_2 for 4 h. (A) The center dark field image of a magnified swirl-type APB. (B) The high-resolution lattice image of a swirl-type APB marked by a red rectangle in (A), and (C) the diffraction pattern derived from the fast-Fourier transform (FFT) of lattice image in (B). *From Yang, C.W., Fang, T.T., "Structures and Development Mechanism of the Anti-Phase Boundaries in Sr_2FeMoO_6" 2012. J. Electrochem. Soc. 159 (3), 35.*

in maze-APBs. The proposed mechanism concerning development of APBs in SFMO not only provides the rationale of the simultaneous existence of maze- and swirl-type APBs and their related morphology formation but also can interpret the suggestion that APBs consist mainly of Fe ions, which in turn can significantly affect the magnetic domain structure in the ordered double perovskite.

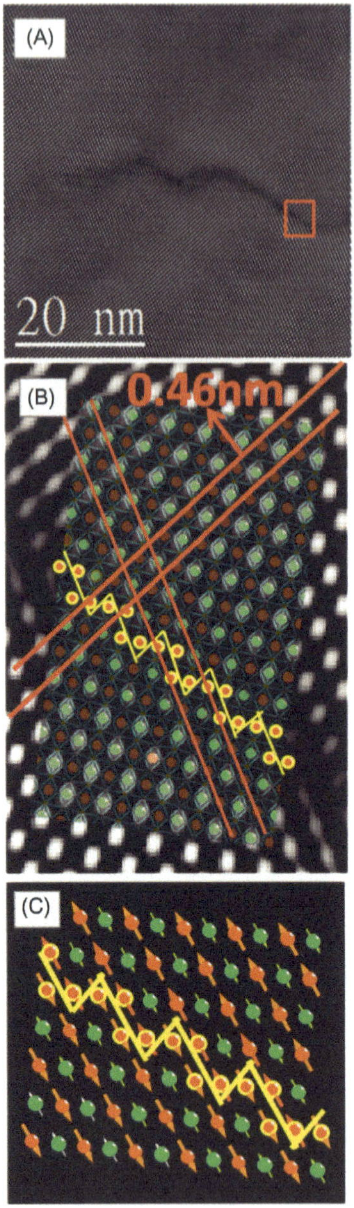

FIGURE 3.28 (A) The simulated B-site lattice image was derived from the inverse-FFT image by masking the supper-lattice spots of (**101**), (**10$\bar{1}$**), ($\bar{1}$**01**), and ($\bar{1}$**0**$\bar{1}$) indicated in Fig. 3.27C. (B) The white spots represent the magnification of the simulated B-site lattice image of the red rectangle marked on (A), whose d-spacing is 0.46 nm consistent with that of {101} planes. By overlaying an ideal lattice of {101} plane of Sr_2FeMoO_6 with B sites indicated by red and green balls, lattice-mismatch sites marked by the red balls encircled by yellow color can be identified along two orthogonal planes of (101) and (**10$\bar{1}$**) (or ($\bar{1}$**0**$\bar{1}$) and ($\bar{1}$**01**)), wherein a zig-zag APB is clearly defined. (C) The schematic diagram of antiferromagnetic configuration of Fe ions on the zig-zag APB shown in (B). *From Yang, C.W., Fang, T.T., Structures and Development Mechanism of the Anti-Phase Boundaries in Sr_2FeMoO_6 2012. J. Electrochem. Soc. 159 (3), 35.*

3.5.6.3 Manganite

The so-called hole-doped manganites (Ramirez, 1997) can be derived from the addition of the divalent metals in $LaMnO_3$. The end member $LaMnO_3$, crystallizing in the orthorhombic perovskite structure with symmetry of Pnma at room temperature, actually is very distorted with the octahedra being elongated and tilted, arising from Jahn–Teller effect in B sites. The structures of the hole-doped manganites with the divalent metals (M), Ca, Sr, Ba, and Pb, preserve typical perovskite, stabilized in the range of t values of $0.\ 85 < t < 0.91$. As defined in Chapter 4, the divalent metals are acceptors and the defect compensation can be either ion with oxygen vacancies or electron with holes. For manganites, it in fact has been evidenced that the defect compensation should be the latter, and holes are localized in Mn ions making Mn^{3+} become Mn^{4+}. Thus the ratio La^{3+}/M^{2+} determines the oxidation state of Mn and thus the ratio Mn^{3+}/Mn^{4+}. This corresponds to the number of Mn sites with a single 3d eg orbital occupied. Double exchange describes the situation where these states simultaneously hop via Mn^{4+} ions. The remaining three t_{2g} electrons on each Mn ion sum up to $s = 3/2$ due to Hund's rule coupling and form a "rigid background."

The influence of the structure on the properties can be realized in terms of size effect of A sites and Jahn–Teller effect of B sites. Concerning the former, the band width and charge transport solely given by e_g states dependent on the size of A-site ions and the spin and orbital orientation of the exchange partners, which in turn would have a significant influence on magnetic ground states and orbital/charge ordered structures. These degrees of freedom react rather cooperatively due to the strong interlink of the octahedral in the perovskite structure. Regarding the latter, a large Jahn–Teller effect, which occurs in the Mn^{3+} ions, produces a strong electron–phonon coupling that persists even at densities where a ferromagnetic ground state is observed. It is natural to imagine the existence of small polarons (see Appendix II) in the paramagnetic phase, and it was believed that these polarons lead to the insulating behavior of this regime. In the context of manganites, while the presence of Mn^{3+} and Mn^{4+} ions together with the site–site double-exchange (DE) mechanism appear to capture the essence of phenomenon of colossal magnetoresistance (CMR), the optimal value of CMR appears at a transition from a high-temperature insulating paramagnetic phase to a low-temperature conducting ferromagnetic phase. Considerable experimental and theoretical investigations have recently suggested (Dagotto et al., 2001; Baldinia et al., 2015) that the ground states of manganites are intrinsically inhomogeneous and characterized by the presence of competing phases containing nanoscale/mesoscale domains. The balance between small (atomic)- and large (domain)-scale interactions strongly affects the onset and the magnitude of the MR in a large variety of manganites. In particular, the role of phase separation and percolation transition is magnified when the competing phases show remarkably different magnetoelectric properties.

Pronounced effects are observed in all physical quantities. However, most spectacular is the CMR at the borderline between a ferromagnetic insulating and ferromagnetic metallic phase. It is at the metal/insulator boundary where the tendencies to form coexisting clusters and percolative transitions are the most important. Note, however, that the metallic phase of manganites at sufficiently large hole density seems quite properly described by DE approaches, namely there is a simple relation between the resistivity and the magnetization in the metallic ferromagnetic phase.

3.5.6.4 $LaBa_2Cu_3O_7$ and $YBa_2Cu_3O_{7-x}$

As observed in Fig. 3.29, the hypothetical structure of $LaBa_2Cu_3{}^{3+}O_9$ can be considered to be derived from the $La_3Cu_3{}^{3+}O_9$ by replacing two of La with Ba and removing one oxygen, and then its oxygen content is further reduced to become $LaBa_2Cu_2{}^{2+}Cu^{3+}O_7$. A distinctive characteristic of lanthanum barium cuprate is that approximately one-third of the total copper is trivalent. The additional oxygen vacancies are accommodated in the top- and bottom-most copper planes. There are two copper sites: a distorted square pyramid of Jahn–Teller distorted Cu^{2+}, forming CuO_2 layers joined at the corners and square planar geometry with fourfold coordinated sites resulting in Cu–O "chains" oriented along [100] (that link the layers).

Although $LaBa_2Cu_3O_7$ (LBCO) was the first high-temperature superconductor discovered, its stability and Curie temperature are relatively low in comparison to $YBa_2Cu_3O_{7-x}$ (YBCO), having the same structure as that of LBCO with La being replaced by Y (Smyth, 1988). Thus the variation of the oxygen content and the structure–property relationship are more intensively studied in YBCO. The structure of YBCO can further be considered as an intergrowth structure with layers stacked in the sequence: Cu(1)O–BaO–Cu(2)O2–Y–Cu(2)O2–BaO–Cu(1)O. The development of Cu–O chains along the b axis is significant since the presence of oxygen atoms in these

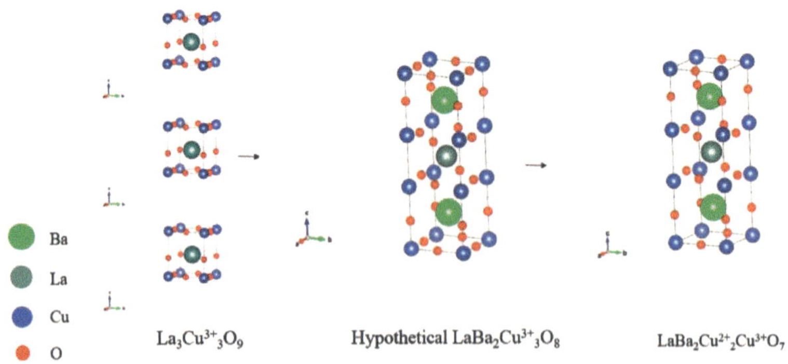

FIGURE 3.29 Origin of the structure of $LaBa_2Cu_3O_7$ considered as a derivative of a perovskite $LaCuO_3$.

chains is essential for superconductivity. The requirement of the integrity of the Cu–O chain is automatic since oxygen vacancies in these chains will drastically distort, and quite probably destroy them. Moreover, the double CuO_2 layers with a remarkable delocalization of the hybridized Cu–O states are a feature common to all the high-temperature superconductors and are believed to be the key to the occurrence of high-temperature superconductivity in these oxides.

The oxygen stoichiometry depending on the copper valence state varies with temperature and oxygen partial pressure. The ordering of the oxygen vacancies along [100] may lead to the change of the crystal symmetry from tetragonal to orthorhombic. Curie temperature (T_c) is strongly dependent on the Cu oxidation state and hence the oxygen content. The average oxidation state of Cu in $YBa_2Cu_3O_{7-x}$ ($0<x<1$) ranges from 2.33 (Cu^{3+}/Cu^{2+}) to 1.66 (Cu^{2+}/Cu^{+}). Underlying the electronic viewpoint, YBCO becomes insulating with a Cu^{2+} (d^9) state in the CuO_2 planes if the oxygen content is lowered to 6. As mentioned in Section 3.4.4, the orbitals of $d_{x^2-y^2}$ for Cu^{2+} (d^9) are singly occupied, and the developing magnetic moments are aligned antiferromagnetically. The Cu^{1+} ions are basically located in the remaining Cu chains, having a closed shell and, therefore, being electronically inert. On the other hand, fully oxidized YBCO has a nominal Cu valence of 2.5 + in the CuO_2 planes and of 2 + in the CuO chains. The Cu^{3+} state was assumed to be present but actually could not be detected, and can be described as a mixture of configurations (Pouchard et al., 2008):

$$a\left|3d^8\right\rangle + b\left|3d^9\underline{L}\right\rangle \tag{3.8}$$

where $b \gg a$ and $\underline{L}$ stands for a hole on the oxygen ligand, where the hole should be equally distributed between Cu^{3+} and oxygen. Due to a coupling between the planes and the chains, holes can be transferred between them, making the hole content in the CuO_2 planes a bit lower than the nominal value, and as a result, the CuO chains become metallic. By internal charge transfer, the d_{z^2} orbital of Cu is occupied in fully oxidized YBCO due to bonding to the apical O_{Ba}. The $d_{x^2-y^2}$ superconducting pair state is stable within a wide range of doping.

3.5.6.5 $CaCu_3Ti_4O_{12}$

The crystal structure of $CaCu_3Ti_4O_{12}$ (CCTO) ($Im\overline{3}$) (Fig. 3.30C) can be derived from the ideal cubic perovskite structure by superimposing a body-centered ordering of Ca and Cu ions and a pronounced titling of the titanium centered octahedral.

Moreover, the significant mismatch in sizes (for instance, for the coordination number of 6, Ca^{2+}: 114 pm versus Cu^{2+}: 87 pm) and bonding preferences of Ca and Cu ions dramatically alter the coordination environments of the A-site cations, leading to a 4-coordinate square-planar environment for

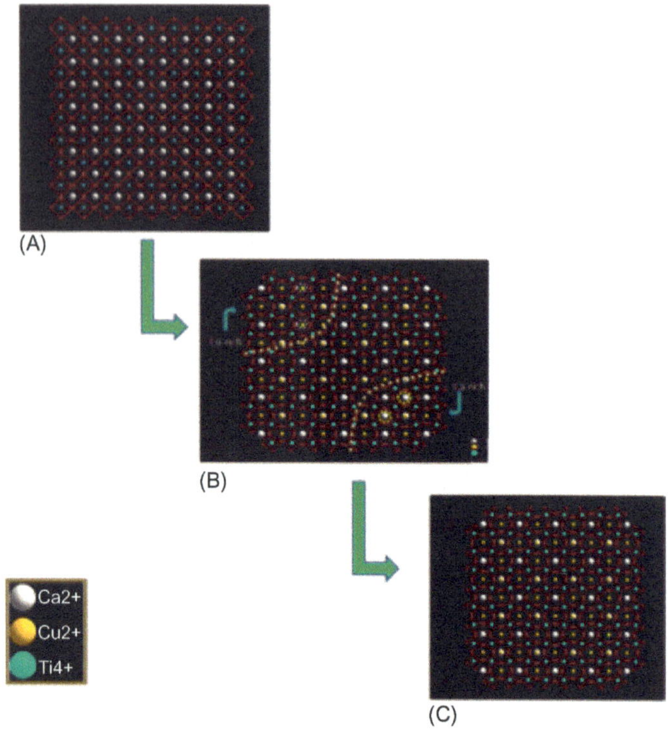

FIGURE 3.30 Schematic structures ({100}): (A) $CaTiO_3$ as a precursor; (B) CCTO with Ca/Cu nanoscale disorders; (C) homogeneous CCTO. *From Fang, T.T., Wang, H.Y., Kuo, J.C., "Role of strained nano-regions in the formation of subgrains in $CaCu_3Ti_4O_{12}$" 2011. J. Appl. Phys. 110, 024103.*

Cu and a 12-coordinate icosahedral environment for Ca. As a consequence, the framework becomes very rigid, and the A-site space is highly restricted, leading to the Ca sites being in a compression state. If so, creation of Ca vacancies would be favorable, like in other well-established CCTO families. However, this is in conflict with the observation of excess Cu species along grain boundaries under the stoichiometric preparation. So far, the widely accepted interpretation concerning the colossal dielectric constant of CCTO is interfacial polarization developed in the grain boundaries, domain boundaries, and the interface between electrodes and materials (Fang and Chung, 2011; Fang and Wang, 2011). While there was controversy concerning the domains, nanobumps due to the lattice distortion have been observed by EBSD (Fang et al., 2011). The understanding of the formation mechanism of nanobumps and evolution of the development of subgrains during thermal etching can be inferred in terms of CCTO structure and the processing-dependent formation of ceramics using solid-state reaction. It is very plausible that it may be energetically favorable for $CaTiO_3$ (Fig. 3.30A) to form at the very beginning of the reaction and act as a precursor phase since it is a stable perovskite structure. Afterward, as Cu ions may diffuse into $CaTiO_3$ to substitute Ca cations to form CCTO, Ca cations would be in a compression state, and thus increase the internal energy, leading to slow reaction kinetics. Therefore, if sintering temperatures are not high enough, excess Cu species may remain along grain boundaries even if the overall composition is taken to be stoichiometric. Meanwhile, nanoscale compositional disorder among Ca/Cu sites (Fig. 3.30B) would naturally occur before homogenization (Fig. 3.30C) is completed. Bump strained nanoregions would develop as a consequence of lattice distortion.

3.5.6.6 *$SrBaNb_2O_6$*

The tungsten bronzes (TBs) can be derived from the idealized cubic perovskite structure by letting the positions of the A-site ions become vacant, and the remaining framework would form the structure of the ReO_3 type. The cubic TBs A_xWO_3 (A = alkali metal, $0.3 \le x \le 0.93$) can further be formed via the partial occupation of the A sites.

Strontium barium niobate $Sr_{1-x}Ba_xNb_2O_6$ (SBN) is a ferroelectric of a tetragonal open TB structure with a solid solution over a wide composition range ($0.25 \leqq x \leqq 0.75$) and partially empty sites at room temperature and shown in Fig. 3.31.

The TB-type structure consists of a network of distorted NbO_6 octahedral sharing corners to form three different types of tunnels parallel to the *c*-axis. Both 12-fold coordinated and ninefold coordinated A sites exist, corresponding to fourfold (A1) and fivefold tunnels (A2), respectively. In SBN, the Ba

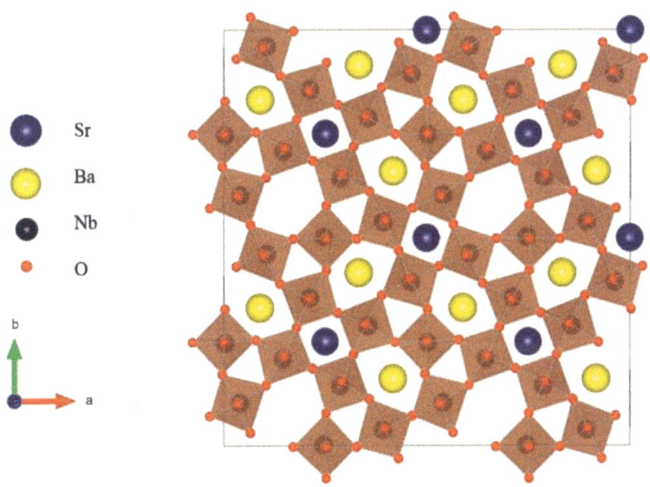

FIGURE 3.31 Structure of $Sr_{1-x}Ba_xNb_2O_6$.

ions prefer the fivefold tunnels and Sr ions the fourfold tunnels, due to strain energy consideration. The NbO_6 octahedra contain two cation (B) sites, M(1) and M(2), and each of them can tilt several degrees in various directions from the *c*-axis. Four of the oxygen ions surrounding each M cation are in the same *z* plane and form a quasi-square. The C-site is the smallest of the three cavities created by the framework of octahedra. The C-site is a tri-capped trigonal prism with nine coordinations (Fang et al., 2010). It was further proposed that strontium and barium ions share the A2 site in roughly a 1 to 3 ratio, while at the A1 site only Sr ions were found. Neither site is fully occupied; the occupancy of each site depends on the Sr/Ba ratio. SBN is a uniaxial relaxor since its polarization is uniquely along the *c*-axis. Relaxor ferroelectrics show enormously large dielectric constants, a pronounced frequency dispersion, and variation of permittivity as a function of temperature. These effects have been attributed to slow relaxation processes and the existence of polar nanoregions (Fang and Chiu, 2011).

SBN has become a very promising material for technological applications and basic research in light of its outstanding properties. Applications include pyroelectric detectors, holographic data storage systems, phase conjugation, generation of photorefractive solitons, quasi-phase-matched second-harmonic generation, and electric-optic modulation.

Part II

Defects of Crystalline Materials

Real crystalline solids are actually never perfect but contain imperfections. Crystal imperfections are not only inevitable features of crystalline solids but also are of primary importance in a great many properties of materials. The dimensionality of imperfections provides a convenient scheme for their classification, which can be identified with one of the following groups: (1) zero-dimensional imperfections, i.e., point defects; (2) one-dimensional imperfections, i.e., line defects; (3) two-dimensional imperfections, i.e., interfacial defects; and (4) three-dimensional imperfections, i.e., bulk defects.

- Point defects, defined as deviations from the perfect atomic arrangement in localized regions of the crystal, are confined to a volume near atomic dimensions including vacant lattice sites, foreign atoms, and extra or missing electrons.
- Line imperfections (dislocations) are defect regions involving large distortions from the average positions along a one-dimensional path, extending through the crystal. The one-dimensional path indicates a space curve, which is not necessarily straight but may be curved or a loop.
- Interfacial defects or surface imperfections are two-dimensional imperfections that break some symmetry of the ordered reference state and serve as boundaries between one orderly region of a crystal and another, which may present as either external (free) surfaces or internal interfaces. The interface is commonly a transition region in which the order of the crystal is interrupted. The order state may refer to the magnetic state, the electron state, the chemical composition, or the configurational arrangement of atoms.
- Bulk defects are microscopic or large scale, which represent inhomogeneity in the shape or structure of the solid including voids, cracks, and inclusions.

Chapter 4

Point Defects in Crystalline Materials

Point defects can be categorized into vacancies, interstitials, and impurities in crystalline solids. But most interstitials are foreign atoms except nonequilibrium defects of self-interstitials developed via irradiation by energetic particles, ion implantation, and cold work. Thus we will treat the interstitials as impurities in this book. Moreover, while these three point defects exist in both metal and ionic solids, there is a principal difference that the defects in the latter are electrically charged. Hence, we will discuss them separately.

4.1 POINT DEFECTS IN METALS

The possible point defects in metals are schematically illustrated in Fig. 4.1.

The vacancies indicate that the lattice sites normally occupied in a perfect crystal are unoccupied, wherein the missing atoms may diffuse to a dislocation core, or a grain boundary, or free surface, or an atom may be forced onto an interstitial site forming the self-interstitials. The impurity defects may be either deliberately doped with dopants or additives, or through contamination. They may form solid solutions, either substitutional or interstitial, or form second phases.

4.1.1 Solid Solutions

A solid solution is considered as a homogeneous mixture of two or more kinds of atoms occurring in the solid state. We refer to the more abundant atomic form as the solvent and to the less abundant atomic form as solute. Two kinds of solid solutions can be classified:

1. Substitutional solid solution: An impurity atom that occupies a normal lattice site.
2. Interstitial solid solution: An impurity atom found in an interstice between matrix atoms.

If two elements can be substituted at all proportions throughout the whole system, they form a continuous or complete series of solid solution. If the solid solubility is limited to only those portions of the phase diagram which

Elements of Structures and Defects of Crystalline Materials.
DOI: http://dx.doi.org/10.1016/B978-0-12-814268-4.00004-7

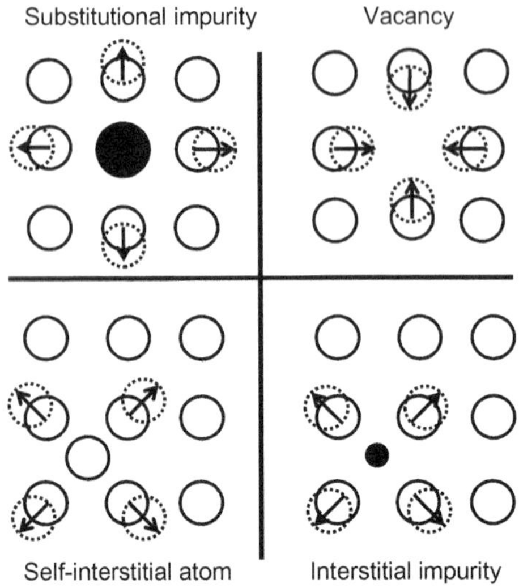

FIGURE 4.1 Illustration of the possible point defects in metals.

are linked to pure elements, the resulting phases are known as primary (or terminal) solid solution.

When an alloying element is added to a given metal in such an amount that the limit of solid solubility is exceeded, a second phase appears with the solid solution. This second phase may be the (primary) solid solution of the alloying element. Such solutions have the same structures as the elements on which they are based; for instance, in the Al–Si system, where the solubility of aluminum for silicon is exceeded, the silicon-rich β-phase appears with the aluminum-rich α-phase. More often the second phase that appears is an intermediate phase whose crystal structure is different from those of the elementary components (pure metal). If these structures occur over a range of composition, they are suggested to be an intermediate solid solution. When the new structures occur with whole-number fixed ratios, the material is referred to as an intermetallic compound with stoichiometric compositions.

The possible intermediate phases can be classified into categories determined by their structures as follows:

1. Interstitial compounds. When the solubility of an interstitially dissolved element is exceeded, an intermediate phase is produced in which these elements are again at interstitial positions, e.g., TaC, TiN, and WC.
2. Valence compounds. Two chemically dissimilar metals tend to form compounds whose bonds show the ionic or covalent binding characteristic, e.g., Mg_2Si, Mg_2Pb, and AlSb.
3. Electron phases. The definite compositions of these phases depend on the ratio of electrons to atoms at those compositions. Consider monovalent

copper (FCC, electron ratio is one electron per atom) added to divalent Zn: the electron ratio increases. At 50 at.%, the electron ratio is three electrons to two atoms, or 1.50 and bcc β-phase (CuZn) appears. At 1.62 (21/13), complex cubic γ-phase (Cu_5Zn_8) appears, and at 1.75 (7/4), an hcp ε-phase ($CuZn_3$) occurs.

4. Order phases. An ordered form of a primary solid solution.
5. Defect phases. A few intermetallic phases contain an unusually large number of vacant lattice sites in certain ranges of concentration and are known as defect phases. NiAl has the CsCl structure. If Al > 50 a/o, the CsCl structure is maintained but vacancies occur in Ni sites (8% at about 54% Al). If Ni > 50 a/o, Ni substitutes Al.

4.1.2 Factors Affecting the Solubility of Impurity Atoms in a Substitutional Solid Solution

For a substitutional solid solution, the solubility is mainly determined by the Hume–Rothery rules:

1. Relative size factor: An extensive solid solubility (on the order of 10%) of one metal in another only occurs if the diameters of the metals differ by less than 15%, related to the strains produced in the lattice of the solvent by the solute atoms. If the size difference is greater than about 15%, solid-solution formation tends to be severely limited and usually is only a fraction of 1%.
2. Chemical affinity factor: The greater the chemical affinity of the two metals, the more restricted is their solid solubility. When their chemical affinity is great, two metals tend to form an intermediate phase rather than a solid solution.
3. Relative valence factor: If the alloying element has a valence different from that of the base metal, the number of valence electrons per atom called the electron ratio, will be changed by alloying. Crystal structures are more sensitive to a decrease in the electron ratio than to an increase. Therefore, a metal of higher valence can dissolve only a small amount of a lower valence metal, whereas the lower valence metal may have good solubility for the high valence metal.
4. Lattice type factor: Only metals that have the same type of lattice can form a complete series of solid solutions. For complete solid solubility, the size factor must be less than 8%.

4.1.3 Factors Affecting the Solubility of the Impurity Atoms in an Interstitial Solid Solution

Regarding an interstitial solid solution, the solubility is dependent on the following two factors:

1. Atomic size: An extensive interstitial solid solution occurs only if the solute atom has an apparent diameter smaller than 0.59 that of the solvent.
2. Electronic structure: Small interstitial solute atoms dissolve much more readily in transition metals than in other metals.

4.1.4 Equilibrium Concentration of Point Defects

To gain a better understanding of the kinetics of material processing, the evaluation of the equilibrium concentration of point defects at a given temperature in crystalline solids shall be taken into account. In adding the point defects to the materials, the lattice energy and entropy of the materials would be changed. Thus equilibrium at constant pressure is defined in terms of the enthalpy (H) and entropy (S) where the system has its minimum value of the Gibbs free energy G,

$$G = H - TS \tag{4.1}$$

Since the main mechanism of the diffusion process is the vacancy mechanism, the equilibrium concentration of vacancy will first be taken into account. Assume that a perfect crystal contains N atoms. If n vacancies are formed, there are two alternative ways to take into account total number sites, i.e., $N+n$ and N. The former indicates that n sites are created by adding the defects but the latter does not. As the vacancies are added in, the free energy of the system will be changed:

$$\Delta G_v = n\Delta h_v - T\Delta S \tag{4.2}$$

where Δh_v is the enthalpy change associated with each vacancy. Upon introducing the vacancies into the crystal, the entropy change is dominated by two parts:

1. A configurational entropy change, ΔS_{conf}, arising from the distribution of n vacancies among the N sites.
2. A vibrational entropy change, ΔS_{vib}, due to the change of frequency in vibration modes of the atoms as vacancies are present.

Thus Eq. (4.2) becomes

$$\Delta G_v = n\Delta h_v - T(\Delta S_{\text{conf}} + \Delta S_{\text{vib}}) \tag{4.3}$$

Concerning the vibrational entropy change due to the presence of vacancies, it can be rationalized as follows. Underlying the statistical mechanics, the Helmholtz free energy of a crystal can be represented by

$$\Delta F = F - F^0 = -kT\sum_i \ln\left(1 - \exp\frac{-h\nu_i}{kT}\right)^{-1} \tag{4.4}$$

where ν_i is the frequency of the ith vibration mode of a crystal. The entropy change can then be obtained by using the thermodynamic relationship

$$\Delta S = -\left(\frac{\partial \Delta F}{\partial T}\right)_V \tag{4.5}$$

for temperatures well above the Debye temperature, i.e., $h\nu_i \ll kT$, and combine Eqs. (4.4) and (4.5), giving

$$\Delta S = -k\sum_i \ln\frac{h\nu_i}{kT} \tag{4.6}$$

Eq. (4.6) basically is not possible to be resolved explicitly in terms of the Debye model. While the Debye model is essentially more realistic in describing the lattice vibration, the calculation of ΔS_{vib} in terms of Debye model is beyond the scope of this book. The related calculation can be referred to (Gallina and Omini, 1964). Nevertheless, for the introduction of a single vacancy into a solid, the vibrational entropy change is local. Thus the Einstein model instead will be taken into account, in which a solid whose crystal lattice structure comprising N atoms can be treated as an assembly of 3N distinguishable one-dimensional quantum harmonic oscillators with all vibration at the same frequency ν. The underlying assumption is that each atom is free to move in three dimensions. For a quantum harmonic oscillator, the Einstein–Bose statistics must be applied.

As a vacancy is introduced, an Einstein model is invoked whereby each atom near the vacancy is considered to vibrate individually as though in the field of its neighbors that are all held fixed. Upon introducing n vacancies in a crystal with coordination number of z, there are nz atoms that have modified frequency ν' and the remaining atoms of N minus nz have the unperturbed frequency ν. Making use of Eq. (4.6), for a perfect crystal,

$$\Delta S_0 = -Nk\ln\frac{h\nu}{kT} \tag{4.7}$$

and for a crystal with n vacancies, the total entropy is

$$\Delta S_{n\nu} = -(N - nz)k\ln\frac{h\nu}{kT} - nzk\ln\frac{h\nu'}{kT} \tag{4.8}$$

The entropy change is given by the difference between Eqs. (4.7) and (4.8):

$$\Delta S_{\text{vib}} = \Delta S_{n\nu} - \Delta S_0 = nzk\ln\frac{\nu}{\nu'} = nk\ln\left(\frac{\nu}{\nu'}\right)^z \tag{4.9}$$

For a single vacancy,

$$\Delta s_{\text{vib}} = \frac{1}{n}\Delta S_{\text{vib}} = k\ln\left(\frac{\nu}{\nu'}\right)^z \tag{4.10}$$

It should be noted that because $\nu' < \nu$,* $\Delta s_{vib} > 0$. Since $\Delta S_{vib} = n\Delta s_{vib}$, Eq. (4.3) can be written as

$$\Delta G_v = n(\Delta h_v - T\Delta s_{vib}) - T\Delta S_{conf} \tag{4.11}$$

Furthermore, ΔS_{conf} is given by Boltzmann's equation:

$$\Delta S_{conf} = k \ln \omega \tag{4.12}$$

where ω is the thermodynamic probability which in this case represents the number of different distinguishable ways for n vacancies distributing on N lattice sites.

Underlying the simple statistical reasoning, it can be shown that the number of different ways of arranging the n vacancies introduced in the N lattice sites is:

$$\omega = \frac{N(N-1)(N-2)\cdots\cdots(N-n+1)}{1\bullet 2\bullet 3\bullet 4\cdots\cdots n} \tag{4.13}$$

Multiply both numerator and denominator of Eq. (4.13) by $(N-n)!$, yielding

$$\omega = \frac{N!}{n!(N-n)!} \tag{4.14}$$

For the case that n vacancies distribute on the $N+n$ lattice sites, it gives

$$\omega = \frac{(N+n)!}{N!n!} \tag{4.15}$$

For a large number of n, Stirling's approximation ($\ln X! = X \ln X - X$, $X \gg 1$) may be applied to Eq. (4.14), and Eq. (4.12) yields

$$\Delta S_{conf} = -k\left(N \ln\frac{N}{N+n} + n \ln\frac{n}{N+n}\right) \tag{4.16}$$

Substitute Eq. (4.16) into Eq. (4.11), giving

$$\Delta G_v = n\Delta h_v - nT\Delta s_{vib} + kT\left(N \ln\frac{N}{N+n} + n \ln\frac{n}{N+n}\right) \tag{4.17}$$

---

* Forces applied on the nearest neighbors of the vacancy change the equilibrium positions of the atoms. Such forces can be internal, e.g., due to electronic redistribution (such as the Jahn-Teller forces), or externally applied forces. In all cases they induce changes in bond lengths in the vicinity of the defect. This results in local modifications of the force constants. When these force constants are lowered, the frequency decreases, due to the fact that the frequency is proportional to the square root of the force constant, and in turn, the response of the lattice becomes softer and at the same time the vibration entropy increases.

At equilibrium,

$$\frac{\partial \Delta G_v}{\partial n} = 0 \tag{4.18}$$

Thus

$$\frac{\partial \Delta G_v}{\partial n} = \Delta h_v - T\Delta s_{\text{vib}} + kT \ln\frac{n}{N+n} = 0 \tag{4.19}$$

Rearrange Eq. (4.19), giving

$$\frac{n}{N+n} = \exp\left(\frac{\Delta s_{\text{vib}}}{k}\right)\exp\left(-\frac{\Delta h_v}{kT}\right) = \exp\left(\frac{T\cdot\Delta s_{\text{vib}} - \Delta h_v}{kT}\right) = \exp\left(-\frac{\Delta g_v}{kT}\right) \tag{4.20}$$

Usually, $N \gg n$, so $N+n \sim N$, and the equilibrium concentration can be expressed as:

$$\frac{n}{N} = \exp\left(-\frac{\Delta g_v}{kT}\right) \tag{4.21}$$

where $\Delta g_v = \Delta h_v - T\Delta s_{\text{vib}}$ is the free energy change associated with the formation of a single vacancy. Note that the entropy term in Eq. (4.20) does not contain the (configurational) entropy of mixing but solely the vibrational entropy change.

With the same argument as we used for vacancy, we can obtain the same form as Eq. (4.11) or (4.17) for a substitutional solid solution:

$$\Delta G_i = n_i(\Delta h_i - T\Delta s_{\text{vib}}) + T\Delta S_{\text{conf}} \tag{4.22}$$

where

$$T\Delta S_{\text{conf}} = kT\left(N \ln\frac{N}{N+n_i} + n_i \ln\frac{n_i}{N+n_i}\right) \tag{4.23}$$

ΔG_i is the free energy change as the impurities added in the system, and Δh_i is the enthalpy change associated with adding each impurity. Note that Δs_{vib} may be positive or negative depending on whether the size of the impurity is smaller or larger than that of the matrix atom, respectively. The equilibrium concentration of a substitutional solid solution can also be derived:

$$\frac{n_i}{N} = \exp\left(-\frac{\Delta g_i}{kT}\right) \tag{4.24}$$

where $\Delta g_i = \Delta h_i - T\Delta s_{\text{vib}}$ is the free energy change associated with each impurity, which can be evaluated by the slope of the graph of ln n_i verse $1/T$ as per Eq. (4.24). According to Eq. (4.24), the number of solute atoms that can be dissolved by the system is zero at 0 K and increases exponentially with increasing temperature. Furthermore, Eq. (4.22) can be represented

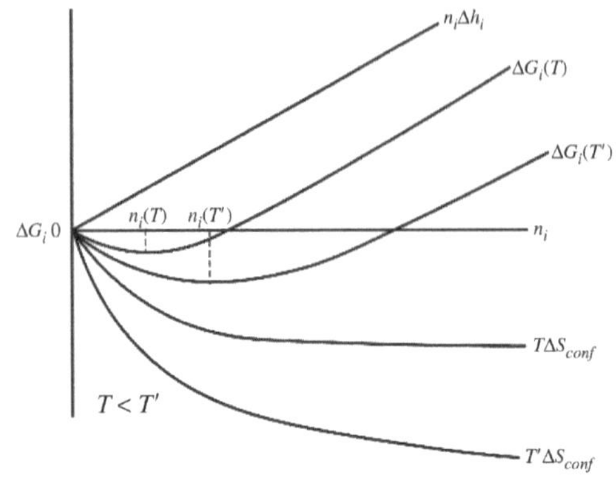

FIGURE 4.2 Changes of free energy, enthalpy, and configurational entropy as a function of the impurity content, n_i, in a crystal at different temperatures.

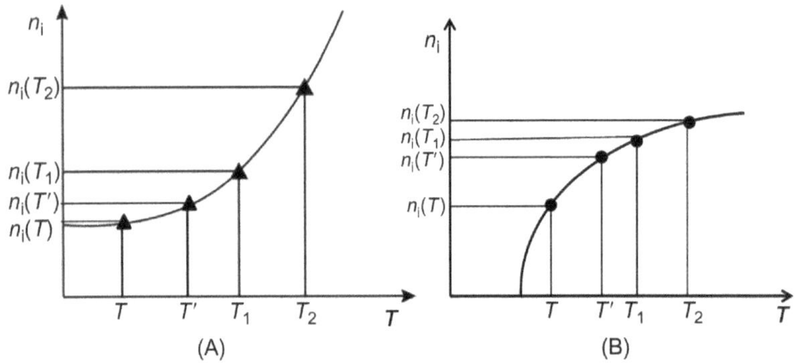

FIGURE 4.3 Equilibrium concentrations of impurity as a function of temperature, where the solid curves as per Eq. (4.24) with (A) high and (B) low free energy change associated with each impurity. T_1 and T_2 are set with the same interval as T and T' for comparison.

graphically in terms of ΔG_i and n_i, shown in Fig. 4.2, wherein since Δs_{vib} is small and can be neglected, $n_i(\Delta h_i - T\Delta s_{\text{vib}}) \approx n_i\Delta h_i$. Moreover, at low pressure, Δh_i is equivalent to the internal energy change associated with each impurity, determined mainly by the strain energy caused by the size difference between the impurity atom and the matrix atom and also insensitive to temperature. On account of a small effect of the temperature on $n_i\Delta h_i$, it reveals only a straight line at different temperatures.

As observed in Fig. 4.2, the minimum in the free energy curve indicates the equilibrium number of solute atoms—an increase as the temperature increases. These two equilibrium numbers of solute atoms at temperatures T and T' are allocated at the theoretical curves as per Eq. (4.24), shown in Fig. 4.3, where the calculated curves reveal different behaviors as the temperature is increased, depending on the free energy change associated with each impurity. The curve shown in Fig. 4.3A indicates that as the temperature is increased, the equilibrium concentration does not change much at low temperatures but increases rapidly at high temperatures. This can be attributed to the high free energy or strain energy change associated with each impurity. Since the strain energy would be highly increased as the impurity dissolves in the matrix at low temperatures, this would lead to the difficulty in dissolving the impurity atoms. But the equilibrium concentration would be enhanced at high temperatures due to the relaxation of the lattice, resulting in low strain energy. By contrast, the curve shown in Fig. 4.3B indicates that the equilibrium concentration is much increased at low temperatures but becomes only slight increased at high temperatures. This can be ascribed to the low free energy or strain energy change associated with each impurity. This is because as the temperature increases, the impurity easily dissolves in the matrix even at low temperatures due to the low strain energy developed. However, the strain energy would become high as the equilibrium concentration reaches a critical amount, leading to the slower increase of the equilibrium concentration at high temperatures. Consequently, it is apparent that the variation of solubility behavior with free energy or strain energy change associated with each impurity is significant in selecting temperatures for completely dissolving the impurity atoms in powder metallurgy and ceramic processing since the incomplete dissolution would have a significant influence on the sintering process. In addition, the inverted diagrams of Fig. 4.3, drawn with T (y axis) versus n_i (x axis), can be referred to as a solubility curve or a solvus line.

The observation of the minima in free energy curve (Fig. 4.2) indicates that the entropy of mixing is always sufficiently large to make the impure system the stable one even if the internal energy of the system increases in the circumstances that solutes are not too high and can be dissolved. This would result in one of the fundamental difficulties in purifying materials due to the fact that they tend naturally to be impure.

For the evaluation of the equilibrium concentration of interstitials, the dissolving of carbon in BCC iron will be set as an example. The concept and procedure of the derivation of the equilibrium concentration of interstitials basically is the same as vacancy. The primary difference is the entropy of mixing. In this system, it involves only the distribution of the carbon atoms in the interstitial sites. Since there are two Fe atoms and six interstitial sites per BCC unit cell, three interstitial sites would be associated with each Fe atom.

For n Fe atoms (n_{Fe}), there are $3n_{\mathrm{Fe}}$ interstitial sites so the entropy of mixing can be expressed as:

$$\Delta S_{\mathrm{conf}} = k \ln \frac{(3n_{\mathrm{Fe}})!}{n_c!(3n_{\mathrm{Fe}} - n_c)!} \tag{4.25}$$

Using the same argument as vacancy, the vibrational entropy change of the existence of interstitials would have the same form as Eq. (4.10):

$$\Delta s_{\text{vib}} = \frac{1}{n}\Delta S_{\text{vib}} = k\,\ln\left(\frac{\nu}{\nu'}\right)^{z'} \tag{4.26}$$

where z' depends on the occupation in octahedral or tetrahedral or combined sites. Since the diameter of carbon atom is about 1.541 nm but the width of the interstitial sites is around 0.0385 nm, the atoms next to the interstitial atom will be pushed much close together, so that $\nu' > \nu$, and thus $\Delta s_{\text{vib}} < 0$. In the same procedure as in deriving the equilibrium concentration of the vacancy, the equation of the equilibrium concentration of the carbon in BCC iron can be derived as follows:

$$\frac{n_c}{(3n_{\text{Fe}} - n_c)} = \exp\left(-\frac{\Delta g_c}{kT}\right) \tag{4.27}$$

because $n_c \ll n_{\text{Fe}}$

$$\frac{n_c}{n_{\text{Fe}}} = 3\,\exp\left(-\frac{\Delta g_c}{kT}\right) \tag{4.28}$$

The theoretical curve concerning the temperature dependence of the equilibrium concentration of the interstitial atoms can be obtained by means of Eq. (4.28).

4.2 POINT DEFECTS IN IONIC SOLIDS

In contrast to metals, the point defects of ionic solids involve different types of ions and charge neutrality. In order to understand the defect structures and the reactions, it is imperative to describe the various defects that can form and to formulate a scheme by which they can be noted. The Kröger–Vink notation is adopted here.

4.2.1 Notation for the Description of Point Defects in Binary Metal Oxides

The point defects are represented by a main symbol along with a subscript and a superscript. The main symbol is the concerned species, the subscript is concerned about the lattice sites or interstitial sites occupied by the species, and the superscript denotes the effective charge of the defect, defined as the difference in charge at the defect site relative to the charge at that site in the perfect crystal. The feature of this notation is described as follows:

Concerning the main symbol and the subscript,

1. In an oxide MO, the metal ions on the normal and regular lattice position are M_M where subscript M describes the type of lattice site which is occupied. The oxygen ions on normal lattice are written as O_O.
2. Vacancies are written V with subscript M or O referring to vacant metal V_M or oxygen V_O site.

3. Interstitial sites indicating the sites are not normally occupied are described with a subscript "*i*," e.g., interstitial oxygen O_i or interstitial metal M_i.
4. An unoccupied or vacant interstitial site is V_i.
5. A foreign cation Mf occupies a regular *M* site (Mf_M) while in an interstitial site (Mf_i).
6. 0: neutral or perfect crystal.

Concerning the effective charge of defects, the related regulations are listed below.

1. Using the perfect crystal as reference, the normal atoms on the regular lattice sites have zero effective charge, and a cross "*X*" is often added as a superscript to symbol M_M^X or O_O^X. For positive and negative effective charge, the symbols of a dot "•" and a prime "′" are used, respectively.
2. The formation of an oxygen vacancy at the oxygen sites. If both of those electrons are localized at the oxygen vacancy, a neutral oxygen vacancy remains and is symbolized as V_O^X. If one or both of the localized or trapped electrons are excited or ionized and transferred away from the neighborhood of the oxygen vacancy, the oxygen vacancy becomes singly or doubly ionized, symbolized as $V_O^{\bullet}$ or $V_O^{\bullet\bullet}$, respectively.
3. For cation vacancies, the symbols V_M^X, $V_M^{'}$, $V_M^{''}$, and $V_M^{'''}$ represent neutral, singly, doubly, and triply ionized metal ion vacancies, respectively, depending on the number of the localized electrons on the cation vacancies.
4. Doubly-charged cation on the interstitial sites: $M_i^{\bullet\bullet}$; doubly-charged oxygen on the interstitial sites: $O_i^{''}$.
5. Substitutional dissolved foreign ions which have a valence different from that of the normal ions are known as aliovalent foreign ions: if the valence is higher than that of the normal metal ion, the foreign ions would possess positive effective charge, e.g., $Mf_M^{\bullet}$. If lower, the foreign ions would possess negative effective charge, e.g., $Mf_M^{''}$.
6. If in the interstitial sites, the foreign cations would possess positive effective charge, e.g., $Mf_i^{\bullet\bullet}$.
7. Quasi-free electrons and holes are denoted as e' and $h^{\bullet}$, respectively.

In a crystal, the neutrality shall always be maintained, namely,

$$\Sigma \text{ positive effective charges} = \Sigma \text{ negative effective charges}$$

In the case of the multivalent cations, like Fe_3O_4, an inverse spinel structure, where Fe^{2+} occupies octahedral sites and Fe^{3+} occupies half of both octahedral and tetrahedral sites, there are two approaches for evaluating the effective charge:

1. Introduce ions with their actual charges and avoid the use of the effective charge.
2. Calculate the effective charge relative to the average actual charge present at the site. For the octahedral sites, this average charge is $(+2(1/2))$,

therefore, the normal lattice constituents have an effective charge $(Fe^{3+}_{Oct})^{(1/2)\bullet}$ or $(Fe^{2+}_{Oct})^{(1/2)'}$. An iron vacancy is represented as $V^{(2(1/2))'}_{Oct}$.

4.2.2 Defect Structures in Stoichiometric Ionic Crystals

An ionic crystal has a stoichiometric composition M_aX_b indicating that it does contain M and X atoms in the exact ratio a/b. In case a point defect is developed and charged in a stoichiometric ionic crystal, an oppositely charged point defect would be created to compensate it in order to conserve the electrical neutrality. Kröger and Vink have proposed six basic types of defect structure for a stoichiometric composition as follows:

1. Cation and anion vacancies ($V''_M + V^{\bullet\bullet}_X$).
2. Vacancies and interstitial ions of the same component $X''_i + V^{\bullet\bullet}_X$ or $V''_M + M^{\bullet\bullet}_i$.
3. Misplaced atoms $X_M + M_X$: This defect does not occur in ionic ceramics, but it has been postulated to occur in covalent ceramics, like SiC.
4. Vacancies and misplaced atoms for the same type of atom ($V_M + M_X$).
5. Interstitial and misplaced atoms ($X_i + M_X$).
6. Interstitial atoms ($X''_i + M^{\bullet\bullet}_i$).

The defect structure of type (1) is called "Schottky" disorder. Formation of Schottky defect pairs can only occur at outer and inner surfaces or dislocations and will diffuse into the crystal until equilibrium is reached. Moreover, Schottky disorder is favored in stoichiometric crystal where the cations and anions are of comparable size, e.g., NaCl. The defect structure of type (2) is known as Frenkel disorder for cation and anti-Frenkel disorder for oxygen. A Frenkel disorder can be formed directly inside the crystal, but Frenkel disorder predominates when the size of the cations and anions are appreciably different.

4.2.3 Defect Structures in Nonstoichiometric Ionic Crystals

The defect structures of nonstoichiometric ionic crystals, taking binary oxides as an example, can be categorized into: oxygen deficiency, MO_{2-y} (or excess metal, $M_{1+x}O_2$) and metal deficiency $M_{1-x}O_2$ (or excess oxygen MO_{2+y}). Predominated defects in oxygen-deficient oxides may be either oxygen vacancies (oxygen deficient), or interstitial cations (excess metal) or both types of defects ($M_{1+y}O_{2-x}$). In metal-deficient oxides the predominated defects may either be metal vacancies (metal deficiency) or interstitial oxygen atom (excess oxygen) or both. These defects would be formed or annihilated by exposing oxide surface to the ambient gas (inert or reducing or oxygen atmosphere) or in conjunction with a higher or lower oxide or metal. Note that both oxygen vacancies and interstitial cations may be compensated by the quasi-free electrons leading to n-type conductors, e.g.,

Nb_2O_5 and TiO_2; whereas, the charged metal vacancies may be compensated by the positive electronic defects (holes) forming p-type conductor, e.g., transition metal oxides like FeO, CoO, and NiO.

4.2.4 Defect Reactions

The defect structures presented in the previous section may all essentially exist in an ionic solid, but in reality, only one or a few will be predominated. Thus it is necessary to establish the defect reaction equations concerning the formation and the mutual interactions of defects, which in turn could evaluate the related defect concentrations. The rules for the formulation of the defect reactions will be described in the following.

4.2.4.1 Preservation of Regular Lattice Sites Ratio

In an ionic crystal, the ratio of the number of the normal cation and anion lattice sites must be preserved as constant and equal to the stoichiometric ratio. This rule indicates that if the lattice site of one of the ions is created or annihilated, the corresponding number of the lattice sites of the other ion also have to be simultaneously created or annihilated in order to keep the ratio of the lattice site of the crystal constant. Thus for an *MX* ionic crystal, the lattice site ratio of *M* and *X* is 1:1, regardless of whether the real composition is stoichiometric or nonstoichiometric. Accordingly, in an M_2X_3 ionic crystal, the ratio of cation and anion lattice sites equals 2:3, and if three anion-vacancy sites are developed via a defect reaction, two cation sites must also be simultaneously created. The general relationship of an M_aX_b ionic crystal can be obtained according to this rule, i.e., the ratio of sum of the number of atoms and vacancies on each sublattice has to be maintained at the stiochiometric ratio or

$$\frac{\mathrm{M}_M + \mathrm{V}_M}{X_X + \mathrm{V}_X} = \frac{a}{b} \tag{4.29}$$

Thus

$$a(X_X + \mathrm{V}_X) = b(\mathrm{M}_M + \mathrm{V}_M) \tag{4.30}$$

and this relationship has to be maintained at all times. Note that this implies that in a defect reaction, the total number of normal lattice sites may change since the ratio of the number of cation and anion sites has to be maintained constant.

4.2.4.2 Mass Balance

In a defect reaction, conservation of the mass has to be maintained, i.e., the number of defects involved must be the same before and after the defect reaction. A vacancy is considered to have zero mass. Electronic defects do not affect the mass balance.

4.2.4.3 *Electroneutrality*

The total effective charge is balanced, namely, it is the same before and after the reaction of the defects.

4.2.5 Formulation of Defect Equations

4.2.5.1 *Stoichiometric Ionic Crystals With Internal Disorder*

In this category, the two most common stoichiometric defects will be described.

4.2.5.1.1 Schottky Disorder

As aforementioned, Schottky disorder involves the existence of equal amounts of cation and anion vacancies. Take the perfect crystal as a reference and consider the created defects of vacancies, and following the rules for the formulation of the defect reactions, the overall equation can be expressed as

$$\mathrm{M}_M^x + X_X^x \rightleftarrows \mathrm{V}_M^{''} + \mathrm{V}_X^{\bullet\bullet} + X_X^x + \mathrm{M}_M^x \tag{4.31}$$

Cancel M and X on both sides in Eq. (4.31), and the net result is:

$$0 \rightleftarrows \mathrm{V}_M^{''} + \mathrm{V}_X^{\bullet\bullet} \tag{4.32}$$

where "0" represents a perfect crystal. Note that we have assumed that the doubly-charged defects predominate and the formation of the Schottky pair results in the formation of two new lattice sites.

4.2.5.1.2 Frenkel Disorder

The defect structure involves equal numbers of vacancy and interstitials of cation. In the same way as formulating the equation of Schottky disorder and assuming doubly-charged defects, the Frenkel defect reaction equation can be written:

$$\mathrm{M}_M^x \rightleftarrows \mathrm{V}_M^{''} + \mathrm{M}_i^{\bullet\bullet} \tag{4.33}$$

Due to the fact that the formation of the Frenkel disorder is the transfer of a cation on the normal site to an interstitial site, no new lattice sites are created in the process. For anti-Frenkel disorder, it involves equal numbers of vacancy and interstitials of anion. The defect equation is written as

$$X_X^x \rightleftarrows X''_i + \mathrm{V}_X^{\bullet\bullet} \tag{4.34}$$

4.2.5.2 *Nonstoichiometric Ionic Crystals*

A nonstoichiometric crystal indicates the presence of an excess or deficit of cations or anions, in which if the predominating types of defects are charged, the compensated electronic defects would be developed for the sake of charge balance. The concept of nonstoichiometry basically can be classified into two types, i.e., intrinsic and extrinsic nonstoichiometry. The former are obtained by the partial oxidation or reduction of a stoichiometric ionic solid and the latter, relevant to the incorporation of foreign ions, is defined as the situation when the aliovalent impurity content exceeds the concentration of any defect arising from intrinsic disorder.

4.2.5.2.1 Intrinsic Nonstoichiometry

4.2.5.2.1.1 Oxygen-Deficient Oxides For an oxide, the change of temperature and oxygen pressure would vary the extent of nonstoichiometry and the defect concentration. For instance, at low oxygen partial pressure, the oxygen atom will escape, the overall reaction may be written

$$MO \rightleftarrows MO_{1-X} + \frac{X}{2}O_2 \tag{4.35}$$

Eq. (4.35) can further be rewritten in terms of defect reaction:

$$O_O^x \rightleftarrows V_O^x + \frac{1}{2}O_2 \tag{4.36}$$

This equation indicates that the escape of the oxygen atom at a normal site would create an oxygen vacancy and leave two electrons. The oxygen vacancy being assigned to be neutral is due to the assumption that the remaining two electrons remain localized at the vacant site. However, they may be excited and freed from the vacancy, which would become singly or doubly charged, depending on the depth of the energy level, the temperature, and vacancy concentration:

$$V_O^x \rightleftarrows V_O^{\bullet} + e' \tag{4.37}$$

$$V_O^{\bullet} \rightleftarrows V_O^{\bullet\bullet} + e' \tag{4.38}$$

Combining Eqs. (4.36), (4.37), and (4.38), gives the net reaction as follows:

$$O_O^x \rightleftarrows V_O^{\bullet\bullet} + \frac{1}{2}O_2 + 2e' \tag{4.39}$$

Concerning the excited electrons,* the status of them depends on the band structure of the crystal. For binary metal oxides, referring to the Appendix II, if the valence electrons of the metals with s or p character, the excited electrons may generally delocalize in the conduction band to form n-type carriers.

* According to the first-principles calculations, the transition level between two charge states of defects may be too deep for electrons to be excited or they may be compensated by the created cation vacancies.

As for the valence electrons of the metals with f character, the excited electrons would localize at the cations on normal sites if the cations would possess negative self-trapping energy (defined in Appendix II). Regarding the valence electrons of the metals with d character, the excited electrons may delocalize in the conduction band or localize at the cations on normal sites if the self-trapping energy of cations is negative, depending on the band structures. It is interesting to point out that for two prototypical materials with identical chemical composition but different crystal lattices, e.g., TiO_2 of rutile and anatase. In rutile, excited electrons can localize at any lattice Ti atom, forming a small polaron, but in anatase, they prefer a band state with a characteristic of large polarons above 60 K. For localized electrons, instead of Eqs. (4.37) and (4.38), the defect reactions may be expressed as

$$M_M^x + V_O^x \rightleftarrows V_O^{\bullet} + M_M' \tag{4.40}$$

$$M_M^x + V_O^{\bullet} \rightleftarrows V_O^{\bullet\bullet} + M_M' \tag{4.41}$$

Consequently, the valence of the cation would be changed, and thus the created electronic defects are often referred to as valence defects.

For metal, excessive oxides with interstitial cation can be an alternative form of oxygen-deficient oxides, the defect reaction may be written as

$$M_M^x + O_O^x \rightleftarrows M_i^x + \frac{1}{2}O_2 \tag{4.42}$$

In this reaction, it is supposed that two vacant sites of the cation and anion have occurred, which, however, would be diffused out to surface or grain boundary. Thus the numbers of cation and anion sites become reduced by one in this reaction. The neutral interstitial cation implies that it gains the electrons from the escaped oxygen ion. Nevertheless, the associated electrons could be successively excited to make the interstitial cation become singly or doubly charged:

$$M_i^x \rightleftarrows M_i^{\bullet} + e' \tag{4.43}$$

$$M_i^{\bullet} \rightleftarrows M_i^{\bullet\bullet} + e' \tag{4.44}$$

The excited electrons may also be ionized to the conduction band or localized at cations.

4.2.5.2.1.2 Metal-Deficient Oxides Consider the cation vacancies as the major defects, which can be easily formed by treating the oxide in a very high oxygen atmosphere via the reaction:

$$\frac{1}{2}O_2 \rightleftarrows V_M^x + O_O^x \tag{4.45}$$

This reaction implies that oxygen is incorporated into the normal site along with the creation of a vacant cation site, indicating that a pair of cation–anion sites is created. In addition, the neutral cation vacancy implies that it associates with two real positive charges or electron holes, which may also be excited. For binary metal oxides, regarding the valence electrons of the metals with s or p character, the excited holes* may generally locate at the valence band of O 2p states to form p-type carriers, which, however, strongly localize at single oxygen ion sites, forming polaron, and become deep lying holes with large effective mass. But the conductivity can be significantly improved, e.g., for the lone pair distorted litharge structure of SnO with the hybridization of O 2p−Sn 5s orbital contribution at the valence band maximum. As for the valence electrons of the metals with d or f character, the excited holes in principle may localize at the cation or oxygen ion sites, depending on the self-trapping energy of cations and oxygen ions. However, the location of the excited holes in NiO is still under debate. In addition, the valence electrons of the metals with d character, the excited holes may become delocalized in the valence band, and the conductivity can be enhanced in the case of the hybridization of d band of cation and 2p band of oxygen that may broaden the valence band and lead to lighter effective masses of holes and hence better carrier mobility. For holes exciting from the cation vacancy, the related defect reaction may be expressed as

$$\mathrm{V}_M^x \rightleftarrows \mathrm{V}_M^{'} + \mathrm{h}^{\bullet} \tag{4.46}$$

The excited holes may delocalize in the valence band, or localize at oxygen ion sites:

$$O_O^x + V_M^x \rightleftarrows V_M^{'} + O_O^{\bullet}$$

or at cation sites:

$$\mathrm{M}_M^x + \mathrm{V}_M^x \rightleftarrows \mathrm{V}_M^{'} + \mathrm{M}_M^{\bullet} \tag{4.47}$$

In such a case, if the cation is +2, its valence will be changed to +3 for $\mathrm{M}_M^{\bullet}$ ions.

The alternative form of metal-deficient oxides is oxygen excess with interstitial oxygen atoms or ions. In this case, the oxygen is incorporated into the interstitial sites, namely,

$$\frac{1}{2}\mathrm{O}_2 \rightleftarrows \mathrm{O}_i^x \tag{4.48}$$

In this reaction, no new sites are created. The neutral oxygen interstitial also associates with two real positive charges or electron holes and may be ionized to the valence band or localized at cations:

$$\mathrm{O}_i^x \rightleftarrows \mathrm{O}_i^{''} + 2\mathrm{h}^{\bullet} \tag{4.49}$$

* According to the first-principles calculations, the transition level between two charge states of defects may be too deep for holes to be excited or they may be compensated by the created oxygen vacancies.

4.2.5.2.1.3 Intrinsic Electronic Reaction As aforementioned, the development of electronic defects is accompanied with the formation of the related point defects. Nevertheless, an intrinsic electron–hole pair can also be formed via the excitation across the bandgap, i.e., an electron is excited from the valence band to the conduction band, leaving an electron hole in the valence band. This reaction is expressed as

$$(e', h^{\bullet}) \text{ or } 0 \rightleftarrows e' + h^{\bullet} \tag{4.50}$$

4.2.5.2.2 Extrinsic Nonstoichiometry: Dissolution of the Aliovalent Impurities in Stoichiometric Ionic Solids

It is well-known that the impurities have a tremendous influence on the material properties, depending on what cations they substitute and what positions they occupy. However, these concerned factors will not be described until Section 4.2.6. Here we will take into account impurity incorporation reactions. Owing to the fact that there are numerous possible alternatives which may occur in different ionic solids, what is attempted here is the illustration of the principles involved in writing the defect reactions. Moreover, in principle, the charge compensation of an aliovalent solute in an ionic solid can be ionic defects (ionically compensated) or electrons and holes (electronically compensated), or a combination of the two (mixed compensation). In the following, several examples will be given in terms of ionic or electronic compensation according to the rules for writing defect equations (Section 4.2.4). Notice that some defect reactions may not be energy feasible, and what the actual defect reactions would be is determined by the experiments. If the respective energies are similar, no single reaction will be the dominant mechanism of incorporation. Nevertheless, one way to look at the energetically feasible alternative is to compare the bandgap and the formation enthalpy for Schottky disorder.

4.2.5.2.2.1 Donors and Acceptors In elemental and compound semiconductors, a dopant per se being able to donate or accept electrons is called a donor or an acceptor, respectively. However, for an ionic solid, it would be more complicated to define the dopant per se as the donor or acceptor since the created defect and vacancy can also donate or accept electrons. Moreover, the charge compensation may be ionic rather than electronic. Assuming that the electrons associated with the defects of either dopants or vacancies can be excited, the related defects would always become positive effective charged; by contrast, regarding holes, the related defects would become negative. Thus it would be convenient to define defects with positive effective charge as donors (D) and negative effective charge as acceptors (A). For example, $V_O^{\bullet}$, $V_O^{\bullet\bullet}$, and $Al_{Mg}^{\bullet}$ are donors but V_{Mg}'', O_i'', and Al_{Ti}' are acceptors. However, it should be kept in mind that it is the precursors in the defect reaction rather than the defined donors or acceptors per se that donate or accept the electrons.

Concerning the energy states of the donors and acceptors for most ionic compounds,* though still located in the bandgap, it would cause confusion in specifying the defect for the energy level regarding each defect reaction since the defects at the same sites possess different energy levels, which basically results from the excitation of electrons or the charge compensation. As the effective charge of the defect is changed by adding or removing an electron in free space, it involves the affinity energy (EA) or ionization energy (IE) of the defect, respectively. The defect in its new valence state is put back to the original lattice site. Thus the overall process is the addition of an electron from outside the crystal to the lattice defect, or the removal of an electron from a lattice defect to outside the crystal, where the related energies are involved in the definition of the positions of occupied and empty levels. In view of the fact that the energy IE for the process $V_O^{\bullet} \rightarrow V_O^{\bullet\bullet} + e'$ is equal to the energy EA for the process $V_O^{\bullet\bullet} + e' \rightarrow V_O^{\bullet}$, the positions of all energy levels can be expressed by the defect reaction involving the IE. In such a manner, the energy level of the defect is labeled with the defect before ionization. Concerning the energy level of oxygen vacancy, V_O^x is labeled instead of $V_O^{\bullet}$ in the following ionization process:

$$V_O^x \rightarrow V_O^{\bullet} + e' \quad \Delta E_1 \tag{4.51}$$

and $V_O^{\bullet}$ is labeled in the following process:

$$V_O^{\bullet} \rightarrow V_O^{\bullet\bullet} + e' \quad \Delta E_2 \tag{4.52}$$

For the energy level of metal vacancy, V_M' and V_M'' are labeled respectively for the following ionization processes:

$$V_M' \rightarrow V_M^x + e' \quad \Delta E_3 \tag{4.53}$$

$$V_M'' \rightarrow V_M' + e' \quad \Delta E_4 \tag{4.54}$$

Regarding dopants, $D_M^{\bullet}$ and A_M' are labeled in the following processes:

$$D_M^{\bullet} \rightarrow D_M^{\bullet\bullet} + e' \quad \Delta E_5 \tag{4.55}$$

$$A_M' \rightarrow A_M^x + e' \quad \Delta E_6 \tag{4.56}$$

The magnitude of the energy levels takes the absolute values of ΔE_i ($i = 1$, 2, 3, 4, 5, and 6). It is worth mentioning that V_O^x is not what we defined as a donor. The energy levels, as labeled, are simply to indicate that they can donate or accept the electrons. The bandgap scheme of MO is schematically represented in Fig. 4.4. Currently, the values of the energy levels of materials have been theoretically calculated based on quantum mechanics. The energy levels near the conduction band are also called donor level (E_D) and near the valence band are called acceptor level (E_A). A deep donor or acceptor is in the

---

* For some transition metal oxides, the valence band may be hybridization of O 2p orbitals with metal 3d orbitals, and the defect energy state may be below the valence band maximum.

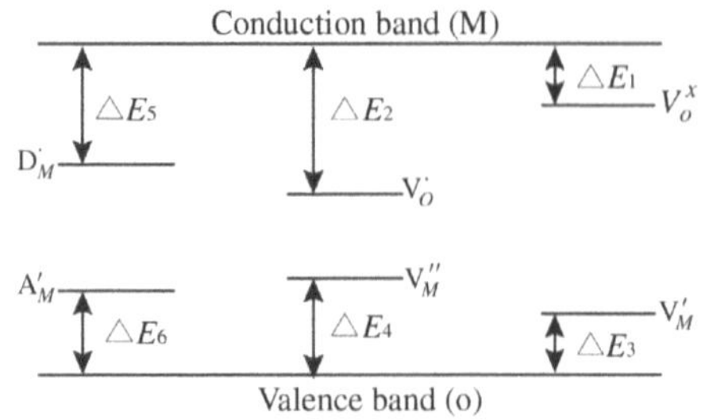

FIGURE 4.4 Energy levels of defects of donors and acceptors in MO.

sense that the energy required to remove an electron or hole from the donor or acceptor level to the conduction or valence band, respectively, is much larger than the characteristic thermal energy kT, where k is the Boltzmann constant and T is temperature. In general, defects with IE of ≤ 0.05 eV are considered shallow. Note that as both donor and acceptor solutes are present, they would compensate for one another, which means that the donor solute contributes its electron to the acceptor solute rather than to the conduction band.

4.2.5.2.2.2 Purely Extrinsic Ionic Disorder or Ionically Compensated Solutes In this case, the relative excess or deficiency of anion is maintained in the crystal and results in the formation of ionic defects.

1. Donor-doped stoichiometric ionic solids
   a. Incorporation of $CaCl_2$ into NaCl

      There are two possible ways to describe the defect reactions of this system:

$$CaCl_2 \xrightarrow{(2NaCl)} Ca_{Na}^{\bullet} + V_{Na}' + 2Cl_{Cl}^x \quad (4.57)$$

$$CaCl_2 \xrightarrow{(NaCl)} Ca_{Na}^{\bullet} + Cl_i' + Cl_{Cl}^x \quad (4.58)$$

      Note that the reaction arrow is pointing only to the right rather than in both directions as was done for the formation of intrinsic disorders, because it was suggested (Smyth, 2000) that the impurities are almost never introduced under equilibrium condition or it is simply ascribed to the fact that the impurities do not diffuse out of the crystal to form the related compound except for the highly volatile materials. In parentheses above the arrow is the matrix compound, explicitly revealing the available lattice sites for the impurity. According to the above two reactions, the positively charged impurity defect, Ca, is shown to be compensated by either the cation vacancy or the anion interstitial. Due to the inadequate space and unfavorable charge environment, the anion interstitial becomes prohibitive. By contrast, there are no such restrictions on the creation of a cation vacancy, which is the favorable choice.

b. Incorporation of Al_2O_3 into MgO

This case basically is the same as (a):

$$Al_2O_3 \xrightarrow{(3MgO)} 2Al^{\bullet}_{Mg} + V''_{Mg} + 3O^x_O \tag{4.59}$$

$$Al_2O_3 \xrightarrow{(2MgO)} 2Al^{\bullet}_{Mg} + O''_i + 2O^x_O \tag{4.60}$$

c. Incorporation of Nb_2O_5 into TiO_2

In this case, there are also two alternative reactions:

$$Nb_2O_5 \xrightarrow{(2TiO_2)} 2Nb^{\bullet}_{Ti} + 4O^x_O + O''_i \tag{4.61}$$

$$2Nb_2O_5 \xrightarrow{(5TiO_2)} 4Nb^{\bullet}_{Ti} + 10O^x_O + V''''_{Ti} \tag{4.62}$$

It should be noted that the anion interstitials (Eq. (4.61)) in a closely-packed anion sublattice are extremely unlikely.

2. Acceptor-doped stoichiometric ionic solids

a. Incorporation of MgO into Al_2O_3

In this case, since the ionic sizes of Mg and Al are relatively close, the energy expenditure for the exchange for one another is small. The Mg ions may enter the solid solution substitutionally (Eq. (4.63)) or interstitially (Eq. (4.64)) or via self-compensation (Eq. (4.65)):

$$2MgO \xrightarrow{(Al_2O_3)} 2Mg'_{Al} + 2O^x_O + V^{\bullet\bullet}_O \tag{4.63}$$

$$3MgO \xrightarrow{(Al_2O_3)} 3Mg^{\bullet\bullet}_i + 3O^x_O + 2V'''_{Al} \tag{4.64}$$

$$3MgO \xrightarrow{(Al_2O_3)} 2Mg'_{Al} + Mg^{\bullet\bullet}_i + 3O^x_O \tag{4.65}$$

It is shown that the negatively charged impurity is compensated by the positively charged anion vacancy or impurity interstitial. The impurity into the interstitial sites with positive charge is compensated by the negatively charged cation vacancy of the matrix. These three incorporation mechanisms were evidenced to be favorable.

b. Incorporation of TiO_2 into Nb_2O_5

Three possible defect reactions may be written as follows:

$$2TiO_2 \xrightarrow{(Nb_2O_5)} 2Ti'_{Nb} + 4O^x_O + V^{\bullet\bullet}_O \tag{4.66}$$

$$5TiO_2 \xrightarrow{(2Nb_2O_5)} 4Ti'_{Nb} + 10O^x_O + Ti^{\bullet\bullet\bullet\bullet}_i \tag{4.67}$$

$$5TiO_2 \xrightarrow{(2Nb_2O_5)} 4V'''''_{Nb} + 10O^x_O + 5Ti^{\bullet\bullet\bullet\bullet}_i \tag{4.68}$$

However, Nb_2O_5 has a relatively closely-packed structure, which in general, is not favorable for the interstitial defects. Moreover, the Ti interstitial is quite highly charged, another detrimental factor. Thus, oxygen vacancy would be the preferred defect compensation.

4.2.5.2.2.3 Purely Extrinsic Electronic Disorder or Electronically Compensated Solutes In this case, the excess anion associated with doped crystal is expelled from the structure, leaving behind an equivalent charge or electrons, or the deficiency of the anion is filled from the ambient, creating equivalent holes.

a. Acceptor-doped stoichiometric ionic solids

Incorporation of AO into M_2O_3

The formation of compensating electronic disorder can be written as follows:

$$2AO + 1/2O_2 \xrightarrow{(M_2O_3)} 2A_M' + 3O_O^x + 2h^\bullet \tag{4.69}$$

Here, it will be shown that the above reaction can be derived from the reaction of extrinsic ionic disorder in the following:

$$2AO \xrightarrow{(M_2O_3)} 2A_M' + 2O_O^x + V_O^{\bullet\bullet} \tag{4.70}$$

$$V_O^{\bullet\bullet} + 1/2O_2 \rightleftarrows O_O^x + 2h^\bullet \tag{4.71}$$

Combine Eqs. (4.70) and (4.71), giving

$$2AO + 1/2O_2 \xrightarrow{(M_2O_3)} 2A_M' + 3O_O^x + 2h^\bullet \tag{4.72}$$

In Eq. (4.71), neutral oxygen atoms enter the crystal and become oxygen ions by stealing two electrons from the top of the valence band, the most energetically accessible electrons in the crystal. This leaves two holes in the valence band that replace the positive charge of the lost oxygen vacancies. The same result can also be obtained by the self-compensated extrinsic ionic disorder:

$$3AO \xrightarrow{(M_2O_3)} 2A_M' + A_i^{\bullet\bullet} + 3O_O^x \tag{4.73}$$

Create two stoichiometric sets of M_2O_3 lattice sites:

$$4A_i^{\bullet\bullet} + 3O_2 \rightarrow 4A_M' + 6O_O^x + 12h^\bullet \tag{4.74}$$

Combine Eq. (4.73) and Eq. (4.74) to eliminate $A_i^{\bullet\bullet}$, which can lead to Eq. (4.72).

b. Donor-doped stoichiometric ionic solids

Incorporation of D_2O_5 into MO_2

The electronically-compensated defect reaction can be depicted as:

$$D_2O_5 \xrightarrow{(2MO_2)} 2D_M^\bullet + 4O_O^x + \frac{1}{2}O_{2(g)} + 2e' \tag{4.75}$$

Using the same argument as (a), the above reaction can also be derived from extrinsic ionic disorder. Note that the advantage of viewing the process as the loss or gain of oxygen from the case involving the anion defect is that there is no need for a loss or gain of lattice sites. Thus the following reaction is first taken into account:

$$D_2O_5 \xrightarrow{(2MO_2)} 2D_M^{\bullet} + 4O_O^x + O_i^{''} \tag{4.76}$$

$$O_i^{''} \rightleftarrows V_i + \frac{1}{2}O_{2(g)} + 2e' \tag{4.77}$$

This reaction indicates that the excess interstitial oxygen is expelled from the crystal as a neutral oxygen molecule, it leaves behind electrons of equivalent charge.

Combining Eqs. (4.76) and (4.77), yields

$$D_2O_5 \xrightarrow{(2MO_2)} 2D_M^{\bullet} + 4O_O^x + \frac{1}{2}O_{2(g)} + 2e' \tag{4.78}$$

The material now is nonstoichiometric since it is not equivalent to the combination of stoichiometric binary constituents any more. Next, take the ionically-compensated reaction into consideration:

$$2D_2O_5 \xrightarrow{(5MO_2)} 4D_M^{\bullet} + 10O_O^x + V_M^{''''} \tag{4.79}$$

Remove a stoichiometric set of MO_2 lattice sites:

$$2O_O^x + V_M^{''''} \rightleftarrows O_2 + 4e' \tag{4.80}$$

Sum Eqs. (4.79) and (4.80), and divide by two, giving Eq. (4.78).

The cause for the aliovalent dopants to be designated as donors or acceptors becomes apparent. The addition of an acceptor oxide (case (a)) would develop compensating holes resulting from the gain of the deficiency of oxygen relative to the host oxide. The addition of a donor oxide (case (b)) would create compensating electrons resulting from loss of the oxygen that is excess relative to the host oxide.

In addition, in order to stress the variations of the property per se, the defect reactions may be in different expressions for the same solute in different oxides. For instance:

The dissolution of Li_2O into ZnO

$$Li_2O + 2e' + \frac{1}{2}O_2 \xrightarrow{(2ZnO)} 2Li_{Zn}^{'} + 2O_O^x \tag{4.81}$$

In this case, reduction of the electron concentration is manifested.

The dissolution of Li_2O into CoO

$$Li_2O + \frac{1}{2}O_2 \xrightarrow{(2CoO)} 2Li_{Co}^{'} + 2h^{\bullet} + 2O_O^x \tag{4.82}$$

This example indicates the enhancement of electron holes.

4.2.5.2.2.4 Issue Concerning the Compensating Defects Concerning this issue, it would be significantly influenced by dopant content, temperature, and oxygen partial pressure. Thus in order to explore the mechanism of charge compensation by dopant, conditions are restricted at slight doping, room temperature, and 1atm. In principle, the simple way to assess the related mechanism of charge compensation is to compare the bandgap energy with the formation energy of Schottky defects. If the former is much larger than the latter, ionic compensation would be dominant, whereas vice versa electronic compensation would be dominant. However, it is now realized that wide-gap materials can be compensated by electrons or holes according to the first-principles calculations (Freysoldt et al. 2014). For example, ZnO at room temperature is readily dopable by electrons, whereas NiO is readily dopable by holes but not by electrons. It should be stressed that the related charge compensation is accompanied by nonstoichiometry: anion deficiency in ZnO, but cation deficiency in NiO. It is also noticed that for ZnO, oxygen vacancy hybridized with Zn interstitial would be the origin causing the enhancement of the concentration of electrons (Kim and Park 2009). It has been reported (Lany et al., 2007) that the material conditions needed for a wide-gap oxide to be compensated by electrons or holes are: (1) Either intrinsic defects, such as vacancies or interstitials, or extrinsic impurities, being able to create electron or hole, should have low formation energies and form shallow donors or acceptors. (2) Ionic compensated defects, like the cation vacancy or the anion vacancy, which would depress the creation of electrons and holes, respectively, should have high formation energies. These two conditions can be further understood by the comparison of the critical Fermi energies $E_F^{n,\mathrm{pin}}$ and $E_F^{p,\mathrm{pin}}$, with the conduction band minimum and the valence band maximum, respectively. At a sufficiently high $E_F^{n,\mathrm{pin}}$, the formation energy of defects, able to suppress the creation of electrons will vanish. These defects will form spontaneously and halt the deliberate n-type doping. Similarly, at a critical $E_F^{p,\mathrm{pin}}$, the defects able to suppress the creation of holes, whose formation energy vanishes, will form spontaneously and halt the deliberate p-type doping. It is suggested that large electron affinities facilitate n-type oxides, small ionization potentials facilitate p-type oxides, and n- or p-type oxides can be further facilitated by increasing the electron or hole densities through extrinsic doping with shallow donors or acceptors, respectively. With respect to the effect of dopant content on the transition of electronic compensation to ionic compensation, the donor-doped $BaTiO_3$ is the most studied. However, the related mechanism so far is not fully understood, which may be complicated by the presence of two metal vacancies, Ba and Ti, and the kinetic problem like slow diffusivity of Ti vacancy. Nevertheless, this subject could be made explicit by taking into account the critical Fermi level pinning. As the dopant content is increased, the Fermi level may approach the critical Fermi level pinning at a critical amount of dopant, which in turn would suppress the electronic compensation but make ionic compensation manifested due to the lowering of the formation energy of cation vacancy. Regarding the effect of oxygen partial pressure on the choice of electronic compensation or ionic compensation at fixed dopant content and temperature, the established Kröger–Vink diagram would provide complete interpretation concerning this issue (see Section 4.2.6.2). Furthermore, if temperature is also varied, measurement of the electrical conductivity at different temperatures and oxygen partial pressure may be a useful approach. Examining the slopes in the log conductivity versus log oxygen partial pressure, the related mechanisms can be identified (Choi and Tuller, 1988).

4.2.5.2.3 Extrinsic Nonstoichiometry: Dissolution of the Aliovalent Impurities in Intrinsic Nonstoichiometric Ionic Solids

4.2.5.2.3.1 Oxygen Deficient Oxides

In this case, it is assumed that oxygen deficiency is due to the escape of oxygen, and the related defect reaction, assuming that doubly-charged vacancies predominate, is

$$O_O^x \rightleftarrows V_O^{\bullet\bullet} + \frac{1}{2}O_2 + 2e' \quad (4.83)$$

In the case where a small amount of donor impurity Mf_2O_3 is dissolved in MO with electronic compensation, the defect reaction is expressed as

$$Mf_2O_3 \xrightarrow{(2MO)} 2Mf_M^{\bullet} + 2e' + 2O_o^x + \frac{1}{2}O_2 \quad (4.84)$$

Based on Eqs. (4.83) and (4.84), the defect reaction representing the decrease of the oxygen vacancies can be obtained by subtracting Eq. (4.83) from Eq. (4.84):

$$V_O^{\bullet\bullet} + Mf_2O_3 \rightarrow 2Mf_M^{\bullet} + 3O_O^x \quad (4.85)$$

Therefore, the dissolution of trivalence Mf-ions in oxygen deficient MO results both in the increase of the electron concentration and the decrease of the concentration of charged oxygen vacancies.

For the acceptor-doped case, assume the dissolution of Mb_2O in MO. The defect reaction may be written as

$$2e' + Mb_2O + \frac{1}{2}O_2 \xrightarrow{(2MO)} 2Mb_M' + 2O_o^x \quad (4.86)$$

This reaction can also combine Eq. (4.83) together to represent the enhancement of oxygen vacancies by summing them up:

$$Mb_2O \xrightarrow{(2MO)} 2Mb_M' + O_o^x + V_O^{\bullet\bullet} \quad (4.87)$$

Thus the dissolution of Mb_2O in oxygen-deficient MO will reduce the electron concentration (or increase the electron hole concentration) and increase the oxygen vacancies.

4.2.5.2.3.2 Metal-Deficient Oxides

In this case, it is assumed that metal deficiency is due to the gain of oxygen, and the related defect reaction, assuming that doubly-charged metal vacancies predominate, is

$$\frac{1}{2}O_2 \rightleftarrows V_M'' + O_O^x + 2h^\bullet \tag{4.88}$$

According to Eq. (4.88), it is apparent that the metal-deficient oxides contain metal vacancies and holes. Thus the donor dopants in this system are to consume the concentration of holes rather than creating electrons, which can be represented by the following reaction by summing Eq. (4.84) with intrinsic electronic reaction:

$$2h^\bullet + Mf_2O_3 \xrightarrow{(2MO)} 2Mf_M^\bullet + 2O_O^x + \frac{1}{2}O_2 \tag{4.89}$$

An alternative reaction can be obtained by the sum of Eq. (4.89) and Eq. (4.88) to indicate the increase of the concentration of the charged metal vacancies as follows:

$$Mf_2O_3 \xrightarrow{(3MO)} 2Mf_M^\bullet + V_M'' + 3O_O^x \tag{4.90}$$

The dissolution of acceptor of Mb_2O in metal-deficient MO will increase the electron hole concentration:

$$Mb_2O \xrightarrow{(2MO)} 2Mb_M' + O_O^x + 2h^\bullet \tag{4.91}$$

Alternative reaction can be obtained by subtracting Eq. (4.91) from Eq. (4.88) to indicate the decrease of the concentration of the charged metal vacancies:

$$Mb_2O + V_M'' \rightarrow 2Mb_M' + O_O^x \tag{4.92}$$

Note that the cations in an oxide may exist in more than one oxidation state, such as the transition metal ions. As long as the energy associated with changing the oxidation state of the cations is not too large, the electron defects can change the oxidation state of the cations. Consider Fe_3O_4 which has an inverse spinel structure with two-thirds of the Fe^{3+} ions and one-third of the Fe^{2+} ions. The defect reaction can be described as follows. Assume metal deficiency, thus metal vacancies and holes are created:

$$\frac{1}{2}O_{2(g)} \rightleftarrows O_O^x + V_{Fe}'' + 2h^\bullet \tag{4.93}$$

If holes localize in the metal sites, the oxidation state of the cations will be changed:

$$2Fe^{2+} + 2h^\bullet \rightleftarrows 2Fe^{3+} \tag{4.94}$$

Combining Eqs. (4.93) and (4.94), gives

$$\frac{1}{2}O_{2(g)} + 2Fe^{2+} \rightleftarrows 2Fe^{3+} + O_O^x + V_{Fe}'' \tag{4.95}$$

This is a general defect equation to indicate that electron holes hop between Fe^{2+} and Fe^{3+}.

4.2.6 Defect Equilibrium in Ionic Solids

Structural elements are defined as the regular atoms on their normal sites or the point defects occupying particular sites, e.g., A_A, V_A, ..., etc. The sum or different combinations of structural elements is called a building unit, e.g., $(A_A + B_A)$. Concerning the chemical potential of the defects in ionic solids, since the number of complementary structure elements cannot be kept constant due to the requirement for a definite site ratio, it is not possible to assign a true chemical potential to a structural element. Nevertheless, this difficulty can be avoided by assigning a virtual thermodynamic potential to each structural element. For example, in an ionic solid with a Schottky defect, the virtual chemical potential of the cation vacancies is given by

$$\eta_{V_M} = \left(\frac{\partial G}{\partial n_{V_M}}\right)_{n_M, n_x, n_{V_x}, P, T,} \tag{4.96}$$

The virtual chemical potential can further be considered as a true chemical potential and is expressed as

$$\eta = \eta^0 + RT \ln a \tag{4.97}$$

However, an undefined constant η^0 present in this equation differs from the true chemical potential. Thus the virtual chemical potentials cannot be determined experimentally. In spite of the presence of the undefined constants in the virtual chemical potential, it would be canceled out during the defect reaction.

4.2.6.1 Equilibrium Concentration of Stoichiometric Oxides

4.2.6.1.1 Schottky Defect

In the same argument as described in Section 4.1.4 for metals, the equilibrium concentration of Schottky defects in an M_aO_b oxide as a function of temperature can be derived, whose defect reaction is expressed as follows:

$$\text{Null (or perfect crystal)} = aV_M^{b-} + bV_O^{a+} \tag{4.98}$$

Assuming there are N_{cat} and N_{an} sites along with n_{cat} and n_{an} vacancies, respectively, in an ionic crystal, the number of ways of distributing the cation vacancies: n_{cat} on $(N_{cat} + n_{cat})$ sites is ω_1 and the number of ways of distributing n_{an} anions vacancies on $(N_{an} + n_{an})$ sites is ω_2,

$$\Delta S_{\text{conf}} = k \ln \omega = k \ln \omega_1 \omega_2 \tag{4.99}$$

$$\omega = \frac{(N_{\text{cat}} + n_{\text{cat}})!(N_{\text{an}} + n_{\text{an}})!}{(N_{\text{cat}})!(n_{\text{cat}})!(N_{\text{an}})!(n_{\text{an}})!} \tag{4.100}$$

Take an MO oxide into account, and thus $(N_{\text{cat}} + n_{\text{cat}})/(N_{\text{an}} + n_{\text{an}}) = 1$. Following an argument as was done for metals, an expression for the equilibrium concentration of Schottky disorders can be obtained:

$$\frac{n_{\text{an}}^{\text{eq}} \cdot n_{\text{cat}}^{\text{eq}}}{(N_{\text{an}} + n_{\text{an}}^{\text{eq}})(n_{\text{cat}}^{\text{eq}} + N_{\text{cat}})} \approx \frac{n_{\text{an}}^{\text{eq}} \cdot n_{\text{cat}}^{\text{eq}}}{N_{\text{an}} N_{\text{cat}}} = \exp\left(-\frac{\Delta h_S + T\Delta s_{S\text{vib}}}{kT}\right)$$

$(\because N_{\text{an}} \gg n_{\text{an}}^{\text{eq}}$ and $N_{\text{cat}} \gg n_{\text{cat}}^{\text{eq}})$

For Schottky defect, $n_{\text{an}}^{\text{eq}} = n_{\text{cat}}^{\text{eq}}$, also $N_{\text{an}} = N_{\text{cat}}$

$$\frac{n_{\text{an}}^{\text{eq}} n_{\text{cat}}^{\text{eq}}}{N_{\text{an}} N_{\text{cat}}} = \left(\frac{n_{\text{an}}^{\text{eq}}}{N_{\text{an}}}\right)^2 = \left(\frac{n_{\text{cat}}^{\text{eq}}}{N_{\text{cat}}}\right)^2$$

$$\text{where } n_{\text{cat}} = \frac{n_{\text{cat}}^{\text{eq}}}{n_{\text{cat}}^{\text{eq}} + N_{\text{cat}}} \approx \frac{n_{\text{cat}}^{\text{eq}}}{N_{\text{cat}}}; \qquad n_{\text{an}} = \frac{n_{\text{an}}^{\text{eq}}}{n_{\text{an}}^{\text{eq}} + N_{\text{an}}} \approx \frac{n_{\text{an}}^{\text{eq}}}{N_{\text{an}}} \tag{4.101}$$

$$n_{\text{an}}^2 = n_{\text{cat}}^2 = \exp\left(\frac{\Delta s_{S\text{vib}}}{k}\right) \exp\left(-\frac{\Delta h_S}{kT}\right)$$

$$n_{\text{an}} = n_{\text{cat}} = \exp\left(\frac{\Delta s_{S\text{vib}}}{2k}\right) \exp\left(-\frac{\Delta h_S}{2kT}\right) = \exp\left(-\frac{\Delta g_S}{2kT}\right)$$

where Δg_S, $\Delta s_{S\text{vib}}$, and Δh_S is the change of Gibbs free energy, entropy, and enthalpy of the formation of a single pair of Schottky disorder, respectively. The superscript "eq" represents the equilibrium condition.

In calculating the equilibrium concentration of defects in ceramic materials, in addition to the statistical approach used in deriving Eq. (4.101), the other approach of chemical mass-action law expressions of the pertinent defect reactions can also be utilized. Consider the defect equilibrium for Schottky defects with the metal and oxygen vacancies being doubly charged in an oxide MO. The defect equation for their formation is given by

$$0 \rightleftarrows V_M'' + V_O^{\bullet\bullet} \tag{4.102}$$

At low defect concentration, the corresponding defect equilibrium may be expressed as

$$k_s = [V_O^{\bullet\bullet}][V_M''] \tag{4.103}$$

where k_s is the equilibrium constant concerning the Schottky defect pair. Note that in the following, square brackets denoting the concentration (mole or site fraction) of defects, and a shorthand notation for the concentration of electrons and holes being n and p, respectively, will be adopted. Essentially, the value of the equilibrium constant is dependent on the units of concentrations employed. If mole fraction is used, k_s is related to the Gibbs molar energy, ΔG_s, of formation of pairs of doubly-charged Schottky defects:

$$k_s = \exp\left(-\frac{\Delta G_S}{RT}\right) \tag{4.104}$$

For equilibrium thermodynamics of the chemical reactions, ΔG_s is equal to the sum of enthalpy change, ΔH_s, and entropy change, ΔS_s. Note that these three terms are in the molar base and the entropy involves vibrational entropy without configurational entropy. Thus

$$\begin{aligned} k_s &= \exp\left(\frac{\Delta S_S}{R}\right)\exp\left(-\frac{\Delta H_S}{RT}\right) \\ \text{Taking } k_0 &= \exp\left(\frac{\Delta S_S}{R}\right) \\ k_s &= k_0 \exp\left(-\frac{\Delta H_S}{RT}\right) \end{aligned} \tag{4.105}$$

Based on the electroneutrality condition, the concentrations of the metal and oxygen vacancies in MO are equal:

$$[V_O^{\bullet\bullet}] = [V_M^{''}] = k_S^{1/2} = k_0^{1/2}\exp\left(-\frac{\Delta H_S}{2kT}\right) \tag{4.106}$$

Eq. (4.106) reveals that the defect concentrations are only dependent on the temperature; they are independent of the activities or the partial pressure of M and O_2.

4.2.6.1.2 Frenkel Defects

Frenkel disorder involves the shift of an ion from its normal lattice site to an interstitial site leaving an empty normal lattice site. Frenkel reaction in an MO oxide is

$$\begin{aligned} M_M^x &= V_M^{''} + M_i^{\bullet\bullet} \\ O_O^x &= O_i^{''} + V_O^{\bullet\bullet} \end{aligned} \tag{4.107}$$

The equilibrium concentration of Frenkel defects as a function of temperature in an MO oxide basically can also be derived in the same way as is done for Schottky defects. However, what we have to consider is the change of the configurational entropy including the number of ways to distribute n_i interstitials on N^* interstitial sites (ω_1) and the number of ways to distribute n_{cat} vacancies on N_t total sites of the cation (ω_2) as follows:

$$\Delta S_{conf} = k \ln \omega_1\omega_2, \tag{4.108}$$

$$\omega_1 = \frac{N^*!}{(N^* - n_i)!(n_i)!}, \qquad \omega_2 = \frac{N_t!}{(N_t - n_{cat})!(n_{cat})!} \tag{4.109}$$

Following the argument used for Schottky disorder, the corresponding temperature dependence of the equilibrium concentration of Frenkel disorder can be obtained:

$$\frac{n_{cat}^{eq} n_i^{eq}}{N_t N^*} \approx \exp\left(-\frac{\Delta g_F}{kT}\right) \tag{4.110}$$

It should be noted that the number of N^* is determined by the crystal structure. For instance, for 1 mol of AgBr (NaCl structure), if the silver ions go into the tetrahedral sites, since the number of interstitial sites is the double of the number of lattice sites $\approx 2N_{Av}$ (Avogadro's number), $N^* \approx 2N_{Av}$. Moreover, if Frenkel defects predominate in the pure stoichiometric compound, the concentrations of interstitial cation and vacancies are then equal, and we obtain:

$$n_{cat}^{eq} = n_i^{eq} \approx A \exp\left(-\frac{\Delta g_F}{2kT}\right) \tag{4.111}$$

where A is a constant and Δg_F is the free energy change associated with the formation of a single pair of Frenkel disorder, respectively. Eq. (4.111) also can be expressed in terms of the chemical mass-action law expressions of the pertinent defect reactions. Assume the defects of Frenkel disorder are doubly charged, the related defect reaction is

$$M_M = V_M'' + M_i^{\bullet\bullet} \tag{4.112}$$

The defect equilibrium with constant k_F can be expressed as:

$$\begin{aligned} &k_F[M_M] = [M_i^{\bullet\bullet}][V_M''] \\ &\text{At low defect concentration} \\ &[M_i^{\bullet\bullet}] \text{ and } [V_M''] \ll [M_M] \approx 1 \\ &[M_i^{\bullet\bullet}][V_M''] = k_F \end{aligned} \tag{4.113}$$

Under the electroneutrality condition, the concentrations of interstitial cation and vacancies are equal, and we obtain:

$$[V_M''] = [M_i^{\bullet\bullet}] = k_F^{1/2} = \exp\left(\frac{\Delta S_F}{2k}\right)\exp\left(-\frac{\Delta H_F}{2kT}\right) = \exp\left(-\frac{\Delta G_F}{2kT}\right) \tag{4.114}$$

Eq. (4.114) also reveals that the defect concentrations are only dependent on the temperature and they are also independent of the activities or the partial pressure of M and O_2.

4.2.6.2 Defect Equilibria in Oxides: The Brouwer Diagram or the Kröger–Vink Diagram

The Brouwer diagram or the Kröger–Vink diagram is a way to tackle the variation of equilibrium defect concentrations with changes in activity of a component of the ionic crystal at constant temperature, usually the oxygen pressure, which is on a logarithmic scale and will be adopted in the following.

4.2.6.2.1 An Oxide MO With Schottky Disorder

Consider the defect equilibrium for Schottky defects with the metal and oxygen vacancies being doubly charged in an oxide MO. The defect equation for their formation is given by

$$0 \rightleftarrows V_M'' + V_O^{\bullet\bullet} \tag{4.115}$$

At low defect concentration, the corresponding defect equilibrium may be expressed as

$$k_S = [V_O^{\bullet\bullet}][V_M''] \tag{4.116}$$

where k_S is the equilibrium constant for the Schottky defect pair. Further consider the intrinsic ionization of electrons:

$$np = K_i \tag{4.117}$$

Charge neutrality is expressed by

$$2[V_O^{\bullet\bullet}] + p = 2[V_M''] + n \tag{4.118}$$

At the stoichiometric composition,

$$[V_O^{\bullet\bullet}] = [V_M''] = k_S^{1/2} \tag{4.119}$$

$$n = p = K_i^{1/2} \tag{4.120}$$

1. Region I: medium-reduction region near stoichiometry

   If the oxygen partial pressure is decreased below the value at the stoichiometric composition, the oxide will lose a certain amount of oxygen in order to stay in equilibrium:

   $$O_O^x \rightleftarrows V_O^{\bullet\bullet} + 2e' + \frac{1}{2}O_2 \tag{4.121}$$

   The mass-action expression is:

   $$[V_O^{\bullet\bullet}]n^2 P_{O_2}^{1/2} = K_{V_O^{\bullet\bullet}} \tag{4.122}$$

   where it is assumed that $[O_O^x] \approx 1$. In this region, while the electron concentration can be increased substantially by reduction, the oxygen vacancy concentration is little affected. Since this region is adjacent to the stoichiometric composition, the concentrations of the intrinsic disorder would still be much higher than those of the electron and hole. Thus the condition of charge neutrality can be approximated by

$$[V_O^{\bullet\bullet}] \approx [V_M''] \approx k_S^{1/2} \tag{4.123}$$

An approximate expression for the electron concentration can be obtained by combining Eqs. (4.122) and (4.123), yielding

$$n \approx K_{V_O^{\bullet\bullet}}^{1/2} k_S^{-1/4} P_{O_2}^{-1/4} \tag{4.124}$$

The hole concentration can be obtained by substitute Eq. (4.117) into Eq. (4.124):

$$p \approx K_{V_O^{\bullet\bullet}}^{-1/2} K_i k_S^{1/4} P_{O_2}^{1/4} \tag{4.125}$$

2. Region II: medium-oxidation region near stoichiometry

If the oxygen partial pressure is increased above the value at the stoichiometric composition, the oxide will gain a certain amount of oxygen in order to stay in equilibrium:

$$\frac{1}{2}O_2 \rightleftarrows V_M'' + O_O^x + 2h^\bullet \tag{4.126}$$

The mass-action expression is

$$[V_M'']p^2 = K_{V_M''} P_{O_2}^{1/2} \tag{4.127}$$

As argued in the reduction region, the oxidation reaction for the oxygen partial pressure near that at the stoichiometric composition can create significant numbers of holes while the concentration of intrinsic defects remains essentially unchanged. Thus the approximate condition of charge neutrality is still represented by Eq. (4.123). An approximate expression for the hole concentration can be obtained by combining Eqs. (4.123) and (4.127), giving

$$p \approx K_{V_M''}^{1/2} k_S^{-1/4} P_{O_2}^{1/4} \tag{4.128}$$

The electron concentration can be obtained by substitute Eq. (4.117) into Eq. (4.128):

$$n \approx K_{V_M''}^{-1/2} K_i k_S^{1/4} P_{O_2}^{-1/4} \tag{4.129}$$

The same dependence of oxygen partial pressure for electrons and holes as was found on both regions I and II indicates that they would lie on the same line in the Kröger–Vink diagram shown in Fig. 4.5.

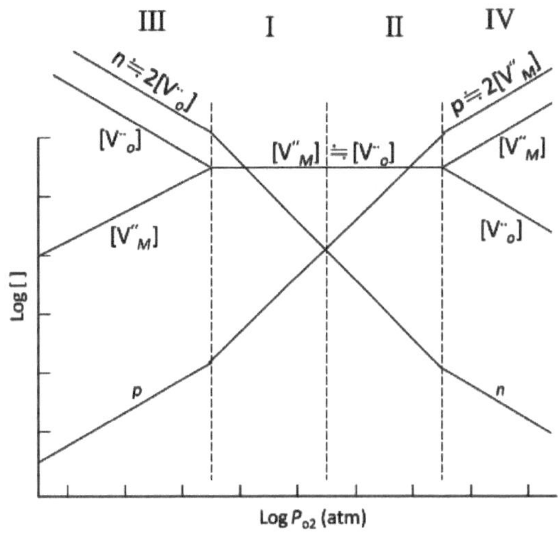

FIGURE 4.5 The Kröger–Vink diagram of MO with doubly-charged oxygen and metal vacancies.

3. Region III: high-reduction region

In this region, the oxygen partial pressure is further decreased, and the reduction reaction becomes the major source of oxygen vacancies and electrons. As described before, the formation of oxygen vacancies is written:

$$O_O^x \rightleftarrows V_O^x + \frac{1}{2}O_{2(g)},$$

$$V_O^x \rightleftarrows V_O^\bullet + e',$$
$$V_O^\bullet \rightleftarrows V_O^{\bullet\bullet} + e',$$

The related mass-action expressions are:

$$[V_O^x]P_{O_2}^{1/2} = k_1[O_O^x] \quad (4.130)$$

$$[V_O^\bullet]n = k_a[V_O^x] \quad (4.131)$$

$$[V_O^{\bullet\bullet}]n = k_b[V_O^\bullet] \quad (4.132)$$

The electroneutrality condition is given by

$$n = [V_O^\bullet] + 2[V_O^{\bullet\bullet}] \quad (4.133)$$

By combining Eqs. (4.130)–(4.133), expressions of the dependence of the oxygen partial pressure of each defect can be obtained.

$$[V_O^{\bullet\bullet}]n^2P_{O_2}^{1/2} = k_1k_ak_b \tag{4.134}$$

$$n = \frac{k_ak_1}{P_{O_2}^{1/2}\cdot n} + 2\frac{k_1k_ak_b}{n^2P_{O_2}^{1/2}} \tag{4.135}$$

$$P_{O_2}^{1/2}n^3 = k_1k_an + 2k_1k_ak_b = k_1k_a(n + 2k_b) \tag{4.136}$$

$$n^3 = k_1k_aP_{O_2}^{-(1/2)}(2k_b + n) \tag{4.137}$$

$$\begin{aligned}&\text{(a) } n \gg 2k_b, \quad n = (k_1k_a)^{1/2}P_{O_2}^{-1/4}\\&\text{(b) } n \ll 2k_b, \quad n = (2k_1k_ak_b)^{1/3}P_{O_2}^{-1/6}\end{aligned}$$

Assume

$$[V_O^x] \gg [V_O^\bullet] + [V_O^{\bullet\bullet}] \qquad [V_O]_t \approx [V_O^x] = k_1P_{O_2}^{-1/2}$$

$$[V_O^\bullet] \gg [V_O^x] + [V_O^{\bullet\bullet}] \qquad [V_O]_t \approx [V_O^\bullet] = (k_1k_a)^{1/2}P_{O_2}^{-1/4}$$

$$[V_O^{\bullet\bullet}] \gg [V_O^x] + [V_O^\bullet] \qquad [V_O]_t \approx [V_O^{\bullet\bullet}] = \left(\frac{1}{4}k_1k_ak_b\right)^{1/3} P_{O_2}^{-1/6}$$

$$[V_O]_t = [V_O^x] + [V_O^\bullet] + [V_O^{\bullet\bullet}]$$

In this region, the dependence of the oxygen partial pressure of electron concentration may range from $P_{O_2}^{-1/4}$ to $P_{O_2}^{-1/6}$ and that of the oxygen vacancy concentration from $P_{O_2}^{-1/2}$ to $P_{O_2}^{-1/6}$. Besides, we can further obtain the dependence of the oxygen partial pressure of metal vacancy and hole concentration from Eq. (4.116) and Eq. (4.117), respectively.

4. Region IV: high-oxidation region

In this region, oxygen partial pressure would become sufficiently high, and it is assumed that the major defects created by the oxidation reaction are metal vacancies and holes. The related defect reactions, mass-action expressions, and the condition of charge neutrality are written as follows:

$$\frac{1}{2}O_2 \rightleftarrows V_M^x + O_O \qquad [V_M^x] = K_3P_{O_2}^{1/2} \tag{4.138}$$

$$V_M^x \rightleftarrows V_M^{'} + h^\bullet \qquad [V_M^{'}]p = K_e[V_M^x] \tag{4.139}$$

$$V_M^{'} \rightleftarrows V_M^{'} + h^\bullet \qquad [V_M^{'}]p = K_f[V_M^{'}] \tag{4.140}$$

$$p = [V_M^{'}] + 2[V_M^{''}] \tag{4.141}$$

Through combinations of the mass-action expressions and electroneutrality above, the concentrations of the ionic and electronic defects can be evaluated. The concentration of electron holes is determined by the expression:

$$p^3 = K_3K_eP_{O_2}^{1/2}[2K_f + p] \tag{4.142}$$

Regarding the electron holes, two limiting conditions may be considered: At (a) low and (b) high concentrations of electron holes,

$$\text{(a)}\ p \ll 2K_f \quad p = (2K_3K_eK_f)^{1/3}P_{O_2}^{1/6}$$

$$\text{(b)}\ p \gg 2K_f \quad p = (K_3K_e)^{1/2}P_{O_2}^{1/4}$$

Following the same argument as in region III, the oxygen pressure dependence of the concentration of the charged metal vacancies can be derived, in which $P_{O_2}^{1/6}$, $P_{O_2}^{1/4}$, and, $P_{O_2}^{1/2}$ for doubly-charged, singly-charged, and neutral metal vacancies, respectively. Besides, we can further obtain the dependence of the oxygen partial pressure of oxygen vacancy and electron concentration from Eq. (4.116) and Eq. (4.117), respectively.

The entire Brouwer or Kröger–Vink diagram for MO based on the above-described regions where only doubly-charged oxygen and metal vacancies are taken into account can now be constructed as shown in Fig. 4.5.

Regarding the region IV with highly oxidized atmosphere, the predominating defect is metal vacancy. However, in some material processing, like sintering, the rate-limiting species may be determined by the concentration of the minority defect, i.e., oxygen vacancy. Thus it would be interesting to illustrate how their concentration varies with that of the major defect. Taking into account the minority defect assumed to be doubly-charged oxygen vacancies, the related mass-action expressions (Eqs. (4.143), (4.144), (4.145), (4.146), and (4.147)) and electroneutrality condition (Eq. (4.148)) are expressed as follows:

$$[V_M^x] = K_3P_{O_2}^{1/2} \tag{4.143}$$

$$[V_M^{'}]p = K_e[V_M^x] \tag{4.144}$$

$$[V_M^{''}]p = K_f[V_M^{'}] \tag{4.145}$$

$$np = K_i \tag{4.146}$$

$$\begin{gathered} O_O^x \rightleftarrows V_O^{\bullet\bullet} + 2e' + \frac{1}{2}O_2 \\ [V_O^{\bullet\bullet}]n^2P_{O_2}^{1/2} = K_{V_O^{\bullet\bullet}} \end{gathered} \tag{4.147}$$

$$2[V_O^{\bullet\bullet}] + p = 2[V_M^{''}] + [V_M^{'}] + n \tag{4.148}$$

Consider the following limiting conditions:

$$(1) \quad [V_M^{''}] \gg [V_M^x] + [V_M^{'}]$$

$$\begin{gathered} \frac{1}{2}O_2 = V_M^{''} + 2h^{\bullet} + O_O^x \\ [V_M^{''}]p^2 = K_{V_M^{''}}P_{O_2}^{1/2} \end{gathered} \tag{4.149}$$

$$p = 2[\mathrm{V}_M'']$$
$$4[\mathrm{V}_M'']^3 = K_{\mathrm{V}_M''} P_{\mathrm{O}_2}^{1/2} \tag{4.150}$$

$$[\mathrm{V}_M''] = \left(\frac{K_{\mathrm{V}_M''} P_{\mathrm{O}_2}^{1/2}}{4}\right)^{1/3} = \left(\frac{K_{\mathrm{V}_M''}}{4}\right)^{1/3} P_{\mathrm{O}_2}^{1/6} \tag{4.151}$$

$$[\mathrm{V}_O^{\bullet\bullet}] = \frac{K_{\mathrm{V}_O^{\bullet\bullet}}}{n^2 P_{\mathrm{O}_2}^{1/2}} = \frac{K_{\mathrm{V}_O^{\bullet\bullet}}}{\left(\frac{K_i}{p}\right)^2 P_{\mathrm{O}_2}^{1/2}} = \frac{K_{\mathrm{V}_O^{\bullet\bullet}} p^2}{K_i^2 P_{\mathrm{O}_2}^{1/2}} = \frac{K_{\mathrm{V}_O^{\bullet\bullet}} (2K_{\mathrm{V}_M''})^{2/3} P_{\mathrm{O}_2}^{1/3}}{K_i^2 P_{\mathrm{O}_2}^{1/2}}$$
$$= \frac{K_{\mathrm{V}_O^{\bullet\bullet}} (2K_{\mathrm{V}_M''})^{2/3}}{K_i^2} P_{\mathrm{O}_2}^{-1/6} \tag{4.152}$$

$$p = 2\left(\frac{K_{\mathrm{V}_M''}}{4}\right)^{1/3} P_{\mathrm{O}_2}^{1/6} \tag{4.153}$$

$$n = \frac{1}{2} K_i \left(\frac{K_{\mathrm{V}_M''}}{4}\right)^{-1/3} P_{\mathrm{O}_2}^{-1/6}$$

Combine Eqs. (4.143), (4.144), and (4.145), giving

$$K_{\mathrm{V}_M''} = K_3 K_e K_f. \tag{4.154}$$

Thus Eq. (4.151) becomes

$$[V_M''] = \left(\frac{1}{4} K_3 K_e K_f\right)^{1/3} P_{\mathrm{O}_2}^{1/6} \tag{4.155}$$

Under this limiting condition, when doubly-charged metal vacancies predominate, the doubly-charged oxygen vacancies decrease with the increase of the oxygen partial pressure, as observed in Fig. 4.6.

(2) $[\mathrm{V}_M'] \gg [\mathrm{V}_M''] + [\mathrm{V}_M^x]$

In the same way as is done in (1), the dependences of the oxygen partial pressure of $[\mathrm{V}_M']$ and $[\mathrm{V}_O^{\bullet\bullet}]$ are given by

$$\frac{1}{2}\mathrm{O}_2 = 2\mathrm{V}_M' + 2\mathrm{h}^\bullet + \mathrm{O}_O^x$$

$$[\mathrm{V}_M']^2 p^2 = K_{\mathrm{V}_M'} P_{\mathrm{O}_2}^{1/2}$$

$$p = [\mathrm{V}_M']$$

$$[V_M'] = (K_3 K_e)^{1/2} P_{\mathrm{O}_2}^{1/4} \tag{4.156}$$

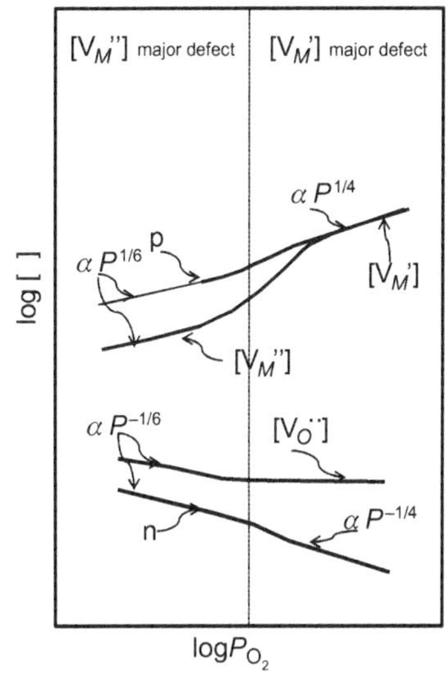

FIGURE 4.6 Schematic presentation of defect concentrations as a function of the oxygen partial pressure in the region IV with highly oxidized atmosphere, in which the predominating defect is metal vacancy and the minority defect is doubly-charged oxygen vacancy.

$$\left[V_O^{\bullet\bullet}\right] = \frac{K_{V_O^{\bullet\bullet}} K_3 K_e}{K_i^2} \tag{4.157}$$

$$n = K_i (K_3 K_e)^{-1/2} P_{O_2}^{-1/4}$$

Under this limiting condition, when singly-charged metal vacancies predominate, the doubly-charged oxygen vacancies become independent of the oxygen partial pressure, shown in Fig. 4.6.

In addition, for a highly reduced region, i.e., region III, in some cases the interstitial cations rather than oxygen vacancies would be more favorable, e.g., the rutile structure TiO_2. It has been reported that these two defects seem to be energetically similar. Thus the simultaneous presence of oxygen vacancies and interstitial metal ions should be taken into account. Assume that doubly-charged oxygen vacancies $V_O^{\bullet\bullet}$ and interstitial cations $M_i^{\bullet\bullet}$ are the major defects in MO_2. The related defect reactions, mass-action expressions, and the condition of charge neutrality are written as follows:

$$O_O^x \rightleftarrows V_O^{\bullet\bullet} + 2e' + \frac{1}{2}O_2 \tag{4.158}$$

$$2O_O^x + M_M^x \rightleftarrows M_i^{\bullet\bullet} + 2e' + O_2 \tag{4.159}$$

$$\left[M_M^x\right] \text{ and } \left[O_O^x\right] \simeq 1$$

$$[V_O^{\bullet\bullet}]n^2P_{O_2}^{1/2} = K_{V_O^{\bullet\bullet}} \quad (4.160)$$

$$[M_i^{\bullet\bullet}]n^2P_{O_2} = K_{M_i^{\bullet\bullet}} \quad (4.161)$$

$$n = 2[V_O^{\bullet\bullet}] + 2[M_i^{\bullet\bullet}] \quad (4.162)$$

Take two limiting conditions into consideration:

a. When $[V_O^{\bullet\bullet}] \gg [M_i^{\bullet\bullet}]$

$$n \simeq 2[V_O^{\bullet\bullet}] \quad (4.163)$$

Combine Eqs. (4.160) and (4.163), yielding

$$\frac{1}{2}n = [V_O^{\bullet\bullet}] = [K_{V_O^{\bullet\bullet}}/4]^{1/3}P_{O_2}^{-1/6} \quad (4.164)$$

By substituting Eq. (4.164) into Eq. (4.161), we obtain

$$[M_i^{\bullet\bullet}] = [K_{M_i^{\bullet\bullet}}/(2K_{V_O^{\bullet\bullet}})^{2/3}]P_{O_2}^{-2/3} \quad (4.165)$$

Under these conditions, the minority concentration of metal interstitials would increase rapidly with decreasing oxygen partial pressure, even more rapidly than the two dominating defects, and may eventually catch up with them and become predominate at the expense of oxygen vacancies.

b. When $[V_O^{\bullet\bullet}] \ll [M_i^{\bullet\bullet}]$

$$n \simeq 2[M_i^{\bullet\bullet}] \quad (4.166)$$

Using the same argument as in (a), the dependences of the oxygen partial pressure of electrons and oxygen vacancies are obtained as follows:

$$\frac{1}{2}n = [M_i^{\bullet\bullet}] = [K_{M_i^{\bullet\bullet}}/4]^{1/3}P_{O_2}^{-1/3} \quad (4.167)$$

$$[V_O^{\bullet\bullet}] = [K_{V_O^{\bullet\bullet}}/(2K_{M_i^{\bullet\bullet}})^{2/3}]P_{O_2}^{1/6} \quad (4.168)$$

It is interesting to point out that the concentration of oxygen vacancies increases with the increase of oxygen partial pressure. The results of the overall analysis are illustrated in Fig. 4.7.

Oxygen sensors are typically conductive ceramics, and operate on one of two principles: One type is based on the Nernst effect, where an electrical potential is created across an oxygen-conducting ceramic, when there is a gradient in oxygen pressure, e.g., ZrO_2. The other type is based on the oxygen pressure dependence of electronic conductivity, e.g., TiO_2.

The above result can be applied to oxygen sensors based on nonstoichiometric TiO_2. Reduction increases the oxygen deficiency x in TiO_{2-x}. This deficiency could be accommodated by oxygen vacancies or titanium

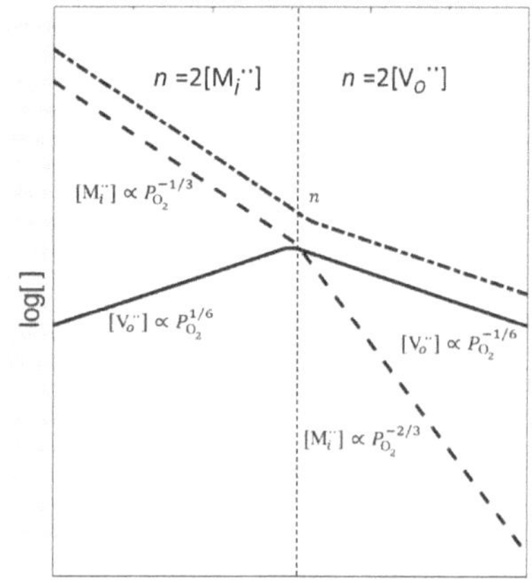

FIGURE 4.7 Schematic presentation of defect concentrations as a function of the oxygen partial pressure in the region III with highly reduced atmosphere, in which the defects of interstitial cations and oxygen vacancies are simultaneously present.

interstitial. The latter has been evidenced as the dominant defect, whose defect reaction and equilibrium are described below.

$$Ti_{Ti}^{x} + 2O_{O}^{x} \rightleftarrows Ti_{i}^{\bullet\bullet\bullet\bullet} + 4e' + O_2 \quad (4.169)$$

Charge neutrality condition is

$$4[Ti_{i}^{\bullet\bullet\bullet\bullet}] = n \quad (4.170)$$

Mass-action expression is

$$K_{Ti_i} = [Ti_{i}^{\bullet\bullet\bullet\bullet}]n^4 P_{O_2} \quad (4.171)$$

Combining Eqs. (4.170) and (4.171), gives

$$n = [4K_{Ti_i}]^{1/5} P_{O_2}^{-1/5} \quad (4.172)$$

The experiment indeed verified that the electron concentration varied with $P_{O_2}^{-1/5}$. However, it is recently evidenced that oxygen vacancy may be the dominant defect in weakly reduced TiO_2, while Ti interstitials can be the majority defect only in strongly reduced anatase, or in case of p-type doping.

4.2.6.2.2 Effects of Impurities on Defect Equilibrium

1. Effect of the cation impurities on the concentrations of the Schottky defects in an ionic crystal.

   Assume that trivalent cation impurity Mf^{3+} is incorporated into an ionic crystal MO with a valence of 2, and thus lead to singly-charged impurity $Mf_{M}^{\bullet}$, where ionic compensation is assumed and intrinsic ionization of electrons is also neglected. The mass-action expressions for the Schottky disorder without impurity are written as:

$$[V_O^{\bullet\bullet}]_0[V_M'']_0 = K_S \tag{4.173}$$

$$[V_O^{\bullet\bullet}]_0 = [V_M'']_0 \tag{4.174}$$

$$[V_M'']_0^2 = K_S \tag{4.175}$$

When impurity is added in, the electroneutrality condition becomes

$$[Mf_M^{\bullet}] + 2[V_O^{\bullet\bullet}] = 2[V_M''] \tag{4.176}$$

By combining the above equations, we obtain

$$[V_M''][V_O^{\bullet\bullet}] = [V_M'']_0^2 = K_S \tag{4.177}$$

$$[Mf_M^{\bullet}] + 2\frac{[V_M'']_0^2}{[V_M'']} = 2[V_M''] \tag{4.178}$$

$$2[V_M'']^2 - [Mf_M^{\bullet}][V_M''] - 2[V_M'']_0^2 = 0 \tag{4.179}$$

$$[V_M''] = \frac{1}{4}\left\{(16[V_M'']_0^2 + [Mf_M^{\bullet}]^2)^{1/2} + [Mf_M^{\bullet}]\right\} \tag{4.180}$$

Take the following limiting conditions into account:

a. When $[Mf_M^{\bullet}] \gg 4[V_M'']_0$

$$[V_M''] = \frac{1}{2}[Mf_M^{\bullet}]$$

Substituting this equation into Eq. (4.177), yields

$$[V_O^{\bullet\bullet}] = 2K_S/[Mf_M^{\bullet}]$$

Under this condition, the concentration of the metal vacancies is determined by that of impurity, called the extrinsic region.

b. When $[Mf_M^{\bullet}] \ll 4[V_M'']_0$

$$[V_M''] = [V_M'']_0$$

In this situation, the concentration of the metal vacancies is not influenced by impurity, called the intrinsic region. These two regions have been represented in Fig. 4.8.

In the same argument as described for Schottky disorders, the effect of donor dopants on the concentration of Frenkel disorders with metal vacancies and interstitials can also be derived.

2. P_{O_2} dependence of the concentrations of the cation impurities doped Schottky defects in an ionic crystal

Take into account an oxide MO containing Schottky defects with double charge as the dominant intrinsic disorder. To construct a Kröger–Vink

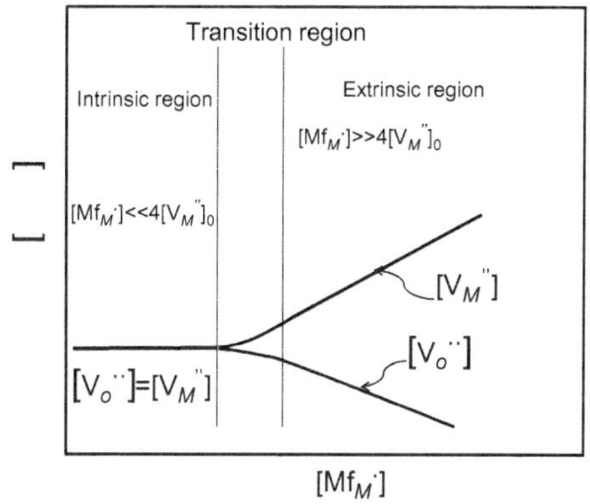

FIGURE 4.8 Effect of higher valent cation impurities on the concentrations of the defects in MO containing mainly Schottky defects.

diagram for doping this material with donor or acceptor impurity, the following essential relationships of the reactions and their mass-action expressions should be considered, and assume that the dopants do not vary the thermodynamic parameters.

a. The intrinsic Schottky disorder

$$0 \rightleftarrows V_M^{''} + V_O^{\bullet\bullet}$$

$$k_S = [V_O^{\bullet\bullet}][V_M^{''}] \tag{4.181}$$

b. The intrinsic electron defects

$$0 \rightleftarrows e' + h^{\bullet}$$

$$np = K_i \tag{4.182}$$

c. The oxidation reaction

$$\frac{1}{2}O_2 = V_M^{''} + 2h^{\bullet} + O_O^x$$

$$[V_M^{''}]p^2 = K_{V_M^{''}} P_{O_2}^{1/2} \tag{4.183}$$

d. The reduction reaction

$$O_O \rightleftarrows V_O^{\bullet\bullet} + 2e' + \frac{1}{2}O_2$$

$$[V_O^{\bullet\bullet}]n^2 P_{O_2}^{1/2} = K_{V_O^{\bullet\bullet}} \tag{4.184}$$

e. Reactions of the donor-doped MO

a. Electronic compensation

$$Mf_2O_3 \xrightarrow{(2MO)} 2Mf_M^{\bullet} + 2e' + 2O_o^x + \frac{1}{2}O_2 \tag{4.185}$$

b. Ionic compensation

$$\mathrm{Mf_2O_3} \xrightarrow{(2\mathrm{MO})} 2\mathrm{Mf}_M^{\bullet} + \mathrm{V}_M^{''} + 3\mathrm{O}_O^{x} \quad (4.186)$$

f. The complete statement of charge balance:

$$2[\mathrm{V}_O^{\bullet\bullet}] + p = 2[V_M^{''}] + n + [\mathrm{Mf}_M^{\bullet}] \quad (4.187)$$

g. Reactions of the acceptor-doped MO

a. Electronic compensation

$$\mathrm{Mb_2O} \xrightarrow{(2\mathrm{MO})} 2\mathrm{Mb}_M^{'} + \mathrm{O}_O^{x} + 2\mathrm{h}^{\bullet} \quad (4.188)$$

b. Ionic compensation

$$\mathrm{Mb_2O} \xrightarrow{(2\mathrm{MO})} 2\mathrm{Mb}_M^{'} + \mathrm{O}_o^{x} + \mathrm{V}_O^{\bullet\bullet} \quad (4.189)$$

h. The complete statement of charge balance

$$2[\mathrm{V}_O^{\bullet\bullet}] + p = 2[V_M^{''}] + n + [\mathrm{Mb}_M^{'}] \quad (4.190)$$

Examining the above relationships, the respective Kröger–Vink diagrams of donor- and acceptor-doped MO can be obtained. For the related comprehensive description refer to related references (e.g., Smyth, 2000). Here, we will illustrate two specific cases in the following.

Case I: Electronic compensation predominated acceptor-doped MO in a highly oxidized atmosphere.

In this case, the hole is the major defect, and cation impurity is taken to be singly-charged $\mathrm{Mb}_M^{'}$. Thus the electroneutrality becomes

$$[\mathrm{Mb}_M^{'}] + 2[\mathrm{V}_M^{''}] = p \quad (4.191)$$

The mass-action expression for the oxidation reaction is

$$[\mathrm{V}_M^{''}]p^2 = K_{\mathrm{V}_M^{''}} P_{\mathrm{O_2}}^{1/2} \quad (4.192)$$

Two limiting conditions will be taken into account:

1. The impurity content is high enough to suppress the concentrations of intrinsic Schottky and electronic disorders, thus

$$[\mathrm{Mb}_M^{'}] \gg [\mathrm{V}_M^{''}]$$

According to Eq. (4.191),

$$[\mathrm{Mb}_M^{'}] \approx p \quad (4.193)$$

Substitute Eq. (4.193) into Eq. (4.192), giving

$$[\mathrm{V}_M^{''}][\mathrm{Mb}_M^{'}]^2 \approx K_{\mathrm{V}_M^{''}} P_{\mathrm{O_2}}^{1/2} \quad (4.194)$$

and

$$[V_M''] \approx \frac{K_{V_M''}}{[Mb_M']^2} P_{O_2}^{1/2} \tag{4.195}$$

2. As the oxygen partial pressure becomes very high, this case would lead to the region IV of Fig. 4.5.

$$2[V_M''] \approx p \tag{4.196}$$

$$p \approx 2^{1/3} K_{V_M''}^{1/3} P_{O_2}^{1/6} \tag{4.197}$$

$$[V_M''] \approx \left(\frac{1}{4}\right)^{1/3} K_{V_M''}^{1/3} P_{O_2}^{1/6} \tag{4.198}$$

The variation of the concentrations of metal vacancies and holes has been schematically illustrated in Fig. 4.9.

Case II: Ionic compensation predominating in acceptor-doped MO_2 in a highly reduced atmosphere.

This case is very significant in the industry of the multilayer ceramic capacitors, where base metal electrodes are used and sintered in a highly reduced atmosphere; and the used capacitor material is, e.g., CaO-doped barium titanate. Thus in this case, an acceptor with double charges is selected, and thus the electroneutrality is expressed as:

$$2[V_O^{\bullet\bullet}] + p = 4[V_M''''] + n + 2[Mb_M''] \tag{4.199}$$

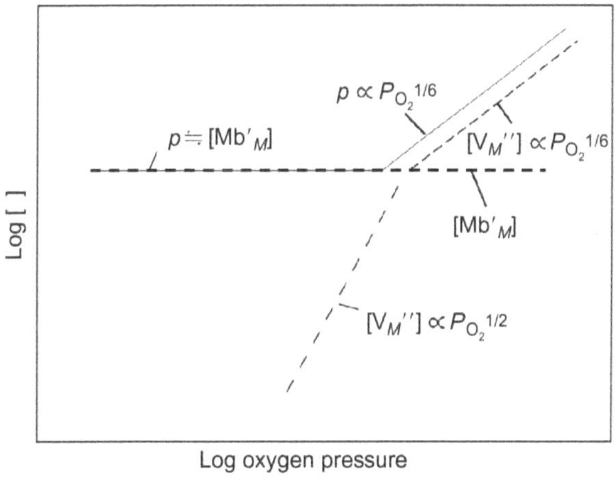

FIGURE 4.9 Schematic presentation of the effect of the oxygen partial pressure on the defect concentrations for electronic compensation predominating in acceptor-doped MO in a highly oxidized atmosphere, wherein the hole is the major defect.

Since this case is in a highly reduced atmosphere, the doubly-charged $V_O^{\bullet\bullet}$ would be the major defect, Eq. (4.199) would be approximated as

$$2[V_O^{\bullet\bullet}] = 2[Mb_M''] + n \tag{4.200}$$

The mass-action expression of reduction reaction is

$$[V_O^{\bullet\bullet}]n^2 P_{O_2}^{1/2} = K_{V_O^{\bullet\bullet}} \tag{4.201}$$

Two limiting conditions will be taken into account:

1. The impurity content is high enough to suppress the concentrations of intrinsic Schottky and electronic disorders. As per Eq. (4.200),

$$[Mb_M''] \approx [V_O^{\bullet\bullet}] \gg n$$

Thus Eq. (4.201) becomes

$$[Mb_M'']n^2 P_{O_2}^{1/2} = K_{V_O^{\bullet\bullet}} \tag{4.202}$$

$$n \approx \frac{(K_{V_O^{\bullet\bullet}})^{1/2}}{[Mb_M'']} P_{O_2}^{-1/4} \tag{4.203}$$

Using the intrinsic electronic equilibrium, we obtain

$$p \approx K_i \frac{[Mb_M'']}{(K_{V_O^{\bullet\bullet}})^{1/2}} P_{O_2}^{1/4} \tag{4.204}$$

It is apparent that the electron concentration would decrease as the content of the acceptor impurity increases under this condition.

2. As the oxygen partial pressure becomes very low, this case would lead to the region III of Fig. 4.5. According to Eq. (4.200),

$$n = 2[V_O^{\bullet\bullet}] \gg 2[Mb_M'']$$

Thus Eq. (4.201) becomes

$$\frac{1}{2} n^3 P_{O_2}^{1/2} = K_{V_O^{\bullet\bullet}} \tag{4.205}$$

$$n = (2K_{V_O^{\bullet\bullet}})^{1/3} P_{O_2}^{-1/6} \tag{4.206}$$

The variation of the concentrations of oxygen vacancies and electrons has been schematically illustrated in Fig. 4.10.

4.2.7 Defect Association

In the previous sections, we have implicitly assumed that the concentration of defects is sufficiently dilute and thus they are totally independent and

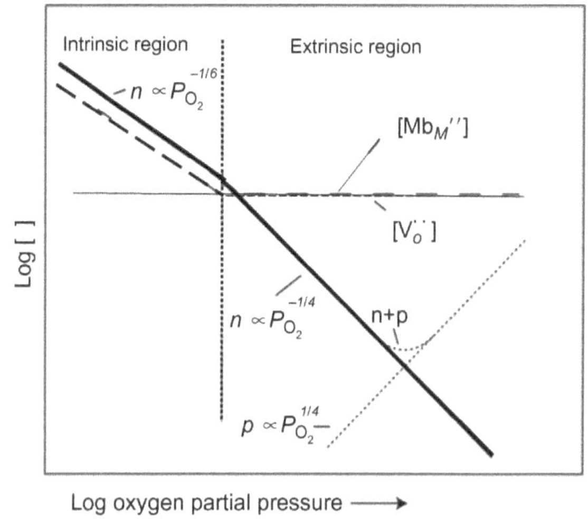

FIGURE 4.10 Schematic presentation of effect of the oxygen partial pressure on the defect concentrations for ionic compensation predominating in acceptor-doped MO_2 in a highly reduced atmosphere, wherein the doubly-charged oxygen vacancy is the major defect.

noninteracting. However, as the concentrations increase, defects will experience a binding energy with other defects, leading to defect associates (combination of two intrinsic ionic defects) or complexes (combination of an impurity with an oppositely charged ionic defect).

In metal, the association of defects is mainly due to the elastic interactions resulting from the local stresses produced by defects, e.g., Ga and As in Si forming donor–acceptor pairs; whereas the formation mechanism of divacancy is dependent on the configuration of the vacancy distribution. For ionic solids, in addition to the strain-induced attraction, the columbic attraction between impurities and point defects would be the main factor to lead to defect association due to the fact that impurities and point defects have opposite effective charges.

Chapter 5

Line Defects in Crystalline Solids

Line defects, i.e., dislocations, play a significant role in plastic deformation and all of the mechanical properties related to plastic deformation, e.g., fracture, creep, and fatigue. There are several reasons why this occurs:

1. Plastic flow of crystals occurs mainly via dislocation motion;
2. Dislocation interactions cause strain hardening; and
3. The kinetics of plastic flow is controlled by the kinetics of dislocation motion.

Nevertheless, this chapter will emphasize the very fundamental characteristics of dislocations without deeply taking into account the plastic deformation processes.

5.1 THE DISCREPANCY BETWEEN THE THEORETICAL AND OBSERVED YIELD STRESSES OF CRYSTALS

In 1926, Frenkel showed that the shear stress for perfect crystal was much higher than the actual shear stresses at which dislocation slip occurred. The former basically refers to the theoretical shear stress necessary to cause a permanent displacement of the atoms, which will now be evaluated in the following. Consider a one-dimensional periodic lattice. As mentioned in Section 2.5 of chapter 2, the atoms in equilibrium positions would lie in the minimum of the related potential well. Since in solids, the atoms are assembled as periodic arrays, the potential energy of the lattice (U_L) can be approximated as shown in Fig. 5.1, in which the lattice positions represent the equilibrium positions with the minimum potential energy and the potential energy at the mid-point is zero indicating the force balance. Underlying the definition concerning the external force (or stress) mentioned in Section 2.5 of chapter 2, the curve of the F_{ext} versus atomic displacement, shown in Fig. 2.3 (dashed line), is redrawn in a periodic lattice and illustrated in Fig. 5.1. As observed, the corresponding force function with atomic position can be assumed to be a sine function in view of the symmetry.

Assuming that a shear stress is applied to a lattice to displace the upper atoms relative to the lower atoms by x, shown in Fig. 5.2, the shear stress may be described as a sine function as aforementioned:

Elements of Structures and Defects of Crystalline Materials.
DOI: http://dx.doi.org/10.1016/B978-0-12-814268-4.00005-9

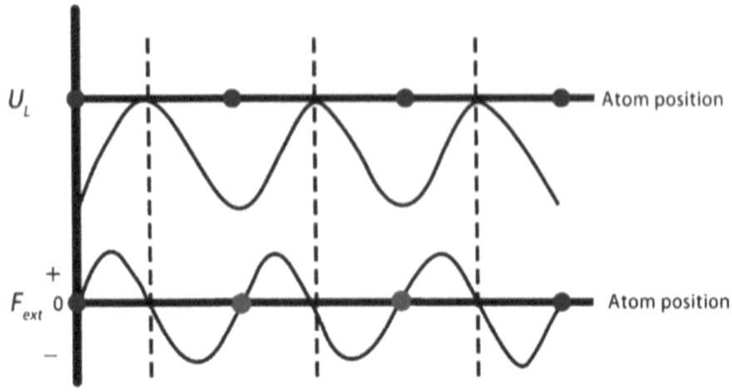

FIGURE 5.1 The lattice potential energy and the corresponding external force function varies with atomic position.

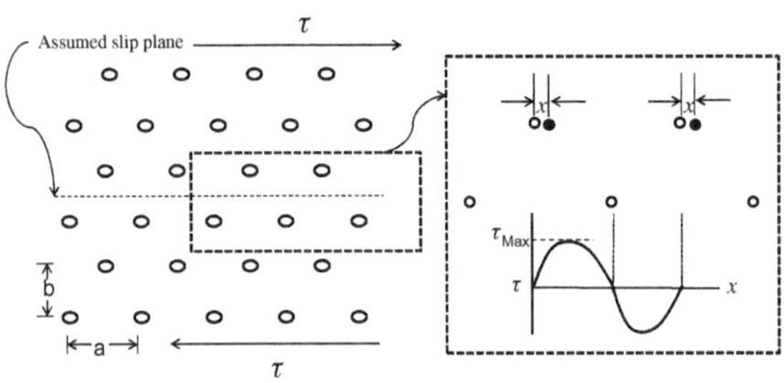

FIGURE 5.2 Illustration of the shear stress function of the displacement as a shear stress is applied on an assumed lattice.

$$\tau = \tau_m \sin\left(2\pi \frac{x}{a}\right) \tag{5.1}$$

At sufficiently small values of displacement, $\sin\left(2\pi \frac{x}{a}\right) \simeq 2\pi \frac{x}{a}$. Thus we may write

$$\tau = \tau_m 2\pi \frac{x}{a} \tag{5.2}$$

According to Fig. 5.2, the shear strain is $\gamma = \frac{x}{b}$. Because Hooke's law also holds at sufficient small displacement, we may write

$$\tau = \mu\gamma = \mu \frac{x}{b} \tag{5.3}$$

In comparing Eq. (5.2) with (5.3), we have

$$\tau_m 2\pi \frac{x}{a} = \mu \frac{x}{b} \tag{5.4}$$

Thus

$$\tau_m = \frac{a}{b} \frac{\mu}{2\pi} \approx \frac{\mu}{2\pi} \simeq \frac{\mu}{10} \tag{5.5}$$

Eq. (5.5) suggests that slip of a perfect lattice should require enormous shear stress $\frac{\mu}{10}$. However, actual shear stress needed for slip typically is 10^{-8}–$10^{-4}\mu$, which is usually referred to as the critical resolved shear stress (CRSS) (Section 5.5). On account of such a considerable difference, it can be concluded that the lattice does not remain perfect during slip, in which some defect such as dislocations must be involved.

5.2 OBSERVATIONS OF DISLOCATIONS

In the past, an overwhelming number of observations have been made which in summary provide unequivocal evidence of the existence of dislocations in crystals. There is such an abundance of these observations that the readers could refer to the related references listed at the end of this book (Hirth and Lothe, 1968). For instance, crystal growth would show a growth spiral associated with a dislocation and etch pits revealed at a free surface is relevant to the strain energy of a dislocation leading to a faster rate of chemical etching. The indirect evidence of dislocations can be observed with the aid of the electron microscope. Fig. 5.3 shows the high resolution electron-transmission micrograph of dislocations in a Sr_2FeMoO_6 double perovskite ceramic material, indicated by an arrow.

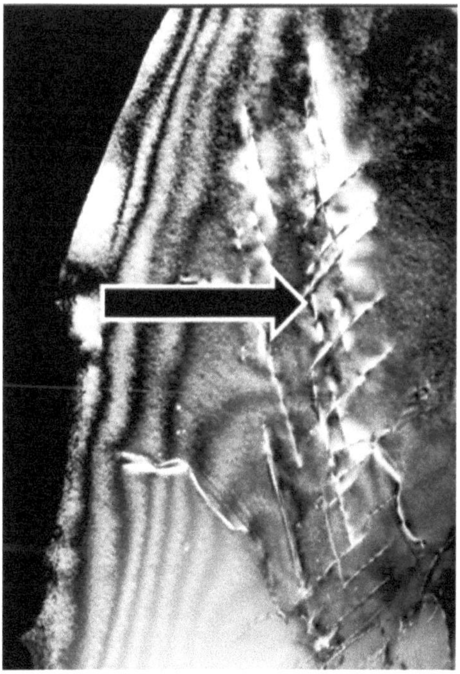

FIGURE 5.3 Evidence of the presence of the dislocations in a Sr_2FeMoO_6 double perovskite ceramic material.

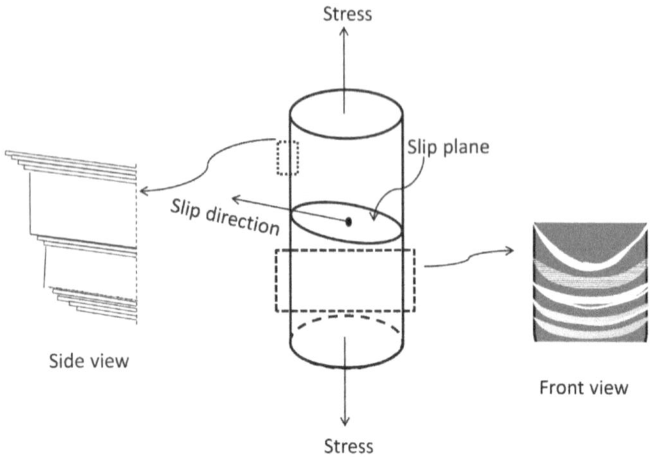

FIGURE 5.4 Magnified schematic view of slip lines of side and front views, appearing on the surface of a cylindrical sample exerted by a stress.

5.3 CRYSTALLOGRAPHIC SLIP

The main experimental evidence of dislocations before the advent of electronic microscopy for observing the crystal structure defects down to the nanometer is slip bands. It was evidenced that the surface of the deformed highly polished single crystal would reveal markings continuously around the specimen. These markings are recognized as the manifestations of a series of microscopic steps that have formed on the surface when viewed by a very high magnification, as shown in Fig. 5.4.

The occurrence of the fine steps suggests that the crystal has undergone "slip" during plastic deformation. The visible markings on the surface are ascribed to slip lines or slip bands, where the crystallographic plane on which the slip has occurred is called the slip plane, and the slip does not occur on just one plane but over several parallel planes. Plastic flow of crystalline materials usually occurs by slip along certain planes and in certain crystal directions. The combination of a slip plane and direction is called a slip system. It is noted that conventionally, the relative translation of one part of a crystal relative to the other along a particular plane and direction, produced by a single dislocation, is called glide, and if by a number of dislocations, is called slip.

5.4 ELEMENTARY GEOMETRIC CHARACTERISTICS OF DISLOCATIONS

Consider a shear stress being exerted on a perfect crystal cube, and assume that the edge dislocation is initiated by slicing the cube from left to right

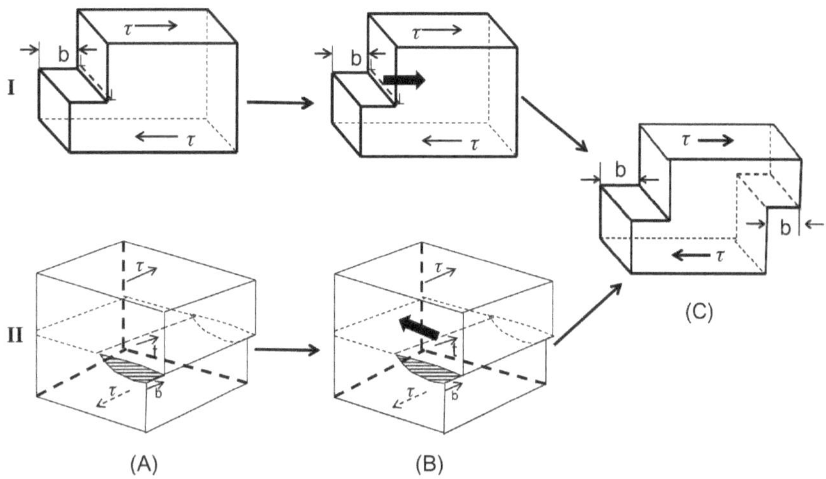

FIGURE 5.5 Path I indicates that (A) a perfect crystal cube is exerted on by a shear stress, by slicing the cube from left to right normal to z and displacing the cube to comply with stress creates an offset with a magnitude of Burgers vector ($\vec{b}$) and the edge dislocation, and (B) the edge dislocation starts to move. Path II indicates that (A) a right-hand screw dislocation is produced by a shear stress where left- and right-hand surfaces simultaneously have an offset with a magnitude of Burgers vector, and (B) the screw dislocation begins to move. (C) The identical shear with a magnitude of Burgers vector, $\vec{b}$, of the crystal cube is developed as either dislocation continues to move through the crystal.

normal to z and displacing the cube to comply with stress as shown in Fig. 5.5A(I), where only the left-hand surface has a displacement with a magnitude of Burgers vector ($\vec{b}$), leading to the creation of an extra half-plane of atoms. Likewise, a right-hand screw dislocation can also initiated through the cut and displacement where left- and right-hand surfaces simultaneously have a displacement with a magnitude of Burgers vector but without an extra half-plane of atoms, as shown in Fig. 5.5A(II). After the initiation of the dislocation, dislocation will move under the action of the stress. However, the moving directions are different, in reference to the stress direction for both dislocations: parallel for the edge dislocation but perpendicular for the screw dislocation. As the dislocation moves through the whole crystal, either dislocation produces the identical shear with a magnitude of Burgers vector, $\vec{b}$, of the crystal cube, as shown in Fig. 5.5C.

Apparently, Burgers vector is the very feature of the dislocations. Burgers vector, also called slip vector, specifies the direction and amount of slip associated with dislocation. The definition of $\vec{b}$ is dependent on the sense of a dislocation line, and the sense is arbitrary, which is defined by assigning a unit vector $\vec{t}$ tangent to the dislocation line and taking the positive sense in

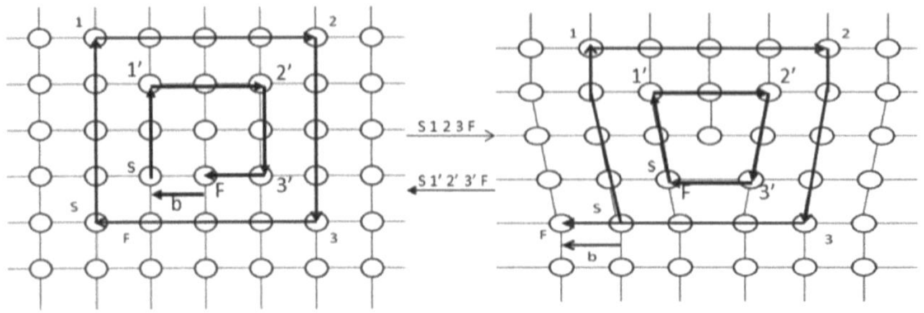

FIGURE 5.6 S123F Burgers circuit represents SF/RH convention: applying an enclosed circuit of a perfect reference crystal to a real crystal containing an extra half plane, where the local Burgers vector is the closure vector. By contrast, S1′2′3′F Burgers circuit represents FS/RH convention: applying an enclosed circuit of a real crystal containing an extra half plane to a perfect reference crystal where the true Burgers vector required to close the circuit.

the positive direction of $\vec{t}$. The procedure to define the Burgers vector by a Burgers circuit illustrated in Fig. 5.6 is as follows:

1. Define a positive sense of $\vec{t}$ being taken to be into the paper.
2. Construct two planes, perfect plane on the left and a plane on the right, containing an edge dislocation perpendicular to the dislocation line.
3. Trace out a clockwise Burgers circuit S123F around the perfect plane on the left, which is enclosed.
4. Applying the same circuit in the plane with a dislocation on the right, reveals the closure vector SF called the local Burgers vector. This convention for b is called the SF/RH convention since the sense of the circuit is that of a right-handed screw RH. The opposite procedure can be applied to define the true Burgers vector, $\vec{b}$. Applying the circuit S1′2′3′F enclosed the plane with a dislocation on the right to the perfect plane on the left. The vector required to close the path from finish F to start S is defined as the true Burgers vector, which is defined as the FS/RH convention.

The local Burgers vector can be further considered as the line integral, taken in a right-handed sense relative to $\vec{t}$, of the elastic displacement around the dislocation. Thus, the local Burgers vector differs than the true Burgers vector, since the former is influenced by the elastic strains and thermal vibrations.

In the same way as applying the Burgers circuit on the front face of a right-handed screw dislocation in a clockwise direction looking down the positive sense of the dislocation line, shown in Fig. 5.7, the Burgers vector as per SF/RH convention is pointing in a positive sense of line. By contrast, the Burgers vector of left-handed screw dislocation points in a negative sense of line. Therefore, the dislocations can be characterized as for an edge dislocation $\vec{b}\cdot\vec{t} = 0$; for a right-handed screw dislocation $\vec{b}\cdot\vec{t} = b$; and for a left-handed screw dislocation $\vec{b}\cdot\vec{t} = -b$.

Accordingly, dislocations can be considered as line imperfections in crystalline solids which are defects that produce distortions centered about a line

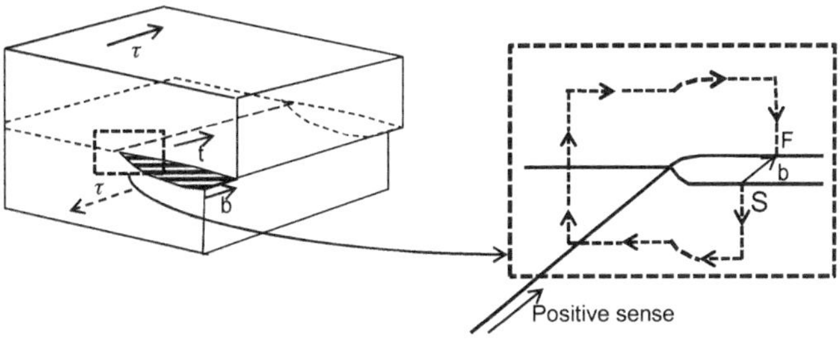

FIGURE 5.7 Schematic presentations of (a) the geometry of the screw dislocation and (b) application of the Burgers circuit on the front face of (a).

and boundary between slipped and unslipped portion of slip plane. Thus dislocations cannot end within a crystal but will either form loops or terminate at a free surface, or an internal surface, like a grain boundary, or end at other dislocations. Note that dislocation loops can have any arbitrary shape.

The characteristics of the fundamental types of dislocations are described in the following.

5.4.1 The Edge Dislocation

1. It may be visualized as an extra half-plane of atoms in a lattice (Fig. 5.6 (right)) at a specified slip plane. If the extra half plane lies above the slip plane, this dislocation is called positive edge dislocation (⊥). If the extra half plane lies below the slip plane, this dislocation is called negative edge dislocation (⊤).
2. The Burgers vector is perpendicular to the sense vector ($\vec{t}$) of the dislocation line.
3. The dislocation line slips in a direction parallel to $\vec{b}$.
4. Since the Burgers vector of an edge dislocation is perpendicular to the dislocation line, its slip plane would be established by both $\vec{b}$ and $\vec{t}$, and thus an edge dislocation can only glide in its own single slip plane.
5. A climb motion is the way for an edge dislocation to move to another slip plane. Reducing the size of the half-plane is called positive climb, accompanied by the annihilation of vacancies, by contrast, increasing its size is called negative climb, accompanied by the generation of vacancies.

The differences between slip and climb are listed as follows:

1. Slip occurs as the result of shear stress, but climb as the result of a normal stress (tensile or compressive).
2. The climb motion of edge or mixed dislocations is nonconservative (involving the annihilation and generation of vacancies) but slip is

conservative. (Conservative motion means that the total number of atoms and lattice sites is conservative.)

3. The climb motion required more energy than slip because of the requirement of vacancy migration, indicating that climb is temperature-dependent but slip is only slightly influenced by temperature.

5.4.2 The Screw Dislocation

1. The geometry of the screw dislocation is illustrated in Fig. 5.7.
2. The Burgers vector is parallel to the dislocation line. Thus any plane containing the screw dislocation is a possible slip plane. However, it should be noted that these planes are restricted to be favorable for the dislocation slip, depending on the structure.
3. The dislocation line moves in a direction perpendicular to $\vec{b}$.
4. No extra half-plane can be visualized in the screw dislocation.

5.4.3 The Mixed Dislocation

Intermediate orientations of Burgers vector and the dislocation line define the mixed character of the dislocations. In Fig. 5.8, one imagines that a curved dislocation line occurs on a slip plane. The unit process of slip, described by the Burgers vector, *b*, has been the same everywhere in the slipped region. Thus the dislocation line has the same Burgers vector at all points along the line. Now we can consider why the atomic displacements adjacent to the dislocation line have various characters depending on the position along the curved line, shown in Fig. 5.8, where near M the dislocation is pure screw, near N it is pure edge, and everywhere else it is mixed.

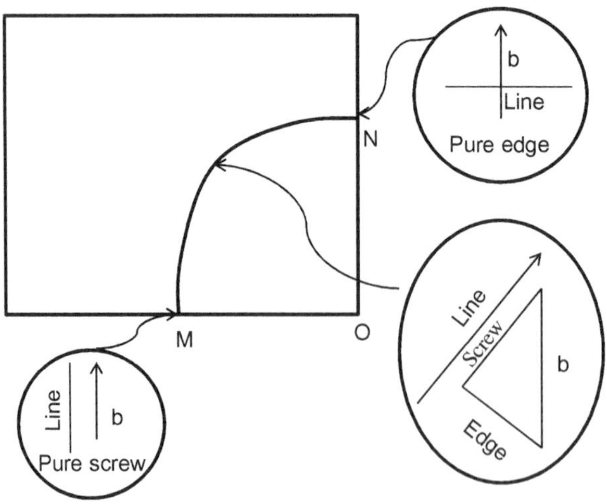

FIGURE 5.8 Schematic illustration of a mixed dislocation on a slip plane.

Summary of the important geometrical properties of dislocations:

1. Dislocations cannot end within a crystal but will either terminate at a grain boundary, or they will form loops, or end at other dislocations. The reason is that a dislocation marks the boundary between the regions where slip has occurred and where it has not, which is the basic characteristic of a dislocation.
2. The Burgers vector is constant along a given dislocation line, resulting from the uniformity of the process of slip within the region bounded by a given dislocation line.
3. When two dislocations, $\vec{b}_1$ and $\vec{b}_2$, combine, the result is a dislocation with Burgers vector $(\vec{b}_1 + \vec{b}_2)$.
4. A dislocation line in direction $\vec{t}$ with Burgers vector $\vec{b}$ defines a slip plane normal by making use of the cross product, i.e., $\vec{b} \times \vec{t}$.

5.5 CRITICAL RESOLVED SHEAR STRESS

In order to produce plastic deformation in polycrystalline or single-crystal metals, a yield stress must be exceeded. Since the plastic deformation is via slip caused by shear stresses, the yield stress would be better expressed in terms of a shear stress resolved on the slip system. Suppose that an axial force F is exerting on a cylindrical crystal of cross-sectional area A, only the component of this force in the slip plane along the slip direction is effective in moving the dislocation, illustrated in Fig. 5.9.

The force component along the slip direction is F cos α, and when it is divided by the area of the slip plane, A/cos β, the corresponding resolved shear stress is obtained:

$$\tau = \frac{F}{A}\cos\alpha\cos\beta = \sigma\cos\alpha\sin\gamma \tag{5.6}$$

where $\sigma = \frac{F}{A}$ is the tensile stress and $\gamma = 90° - \beta$. The term ($\cos\alpha\cos\beta$) has been referred to as the Schmid factor. For any given β, the Schmid factor would have a maximum value at $\alpha = 90° - \beta$. Thus we can obtain the maximum Schmid factor occurring at $\beta = 45°$ with a value of 1/2 after the

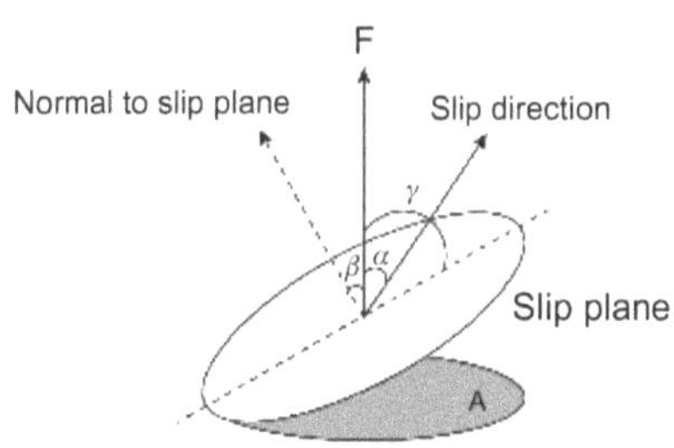

FIGURE 5.9 Geometry for determining resolved shear stress. A is the cross-sectional area of the cylindrical crystal and considered to be the projection area of the slip plane. (β: The angle between the slip plane normal and the applied force axis; α: The angle between the slip direction and the applied force axis).

evaluation of the maximum of the function of $\cos(90-\beta)\cos\beta$. The resolved shear stress required to initiate slip on a given slip system is often called the critical resolved shear stress (CRSS).

The magnitude of CRSS of a single-phase single crystal is varied with the interactions of the dislocations with each other and with defects (such as vacancies and impurity atoms), so CRSS decreases as the density of defects decreases, provided that the total number of imperfections is not zero. At the other extreme, tensile tests on whiskers (nearly free of dislocations) have given strengths which are approximately equal to the calculated strength of a perfect crystal. Under the condition of the presence of a dislocation which does not interact with other dislocations or defects, CRSS can be equivalent to the Peierls stress which is the minimum shear stress required to move a dislocation and can be visualized as an *inherent lattice resistance to the motion of dislocations.*

There are two rules for slip that should be noted:

1. Slip always occurs in close-packed direction, i.e., in the direction of the shortest lattice translation vector (unit of slip).
2. Slip usually occurs on the most widely spaced (or densely packed) planes, but this rule is not always obeyed. For instance, NaCl does not slip on the most densely packed planes {001} but on {110}, however, PbS with the same crystal structure does slip on {001}.

Rigid slip is unrealistic because all atoms in the slip plane are required to slip at the same time. This is unnatural and relative translations always occur gradually.

5.6 PLASTIC FLOW ASSOCIATED WITH THE SLIP OF THE DISLOCATION MOTION

Upon the application of the shear stress, the shear stress will further drive dislocation motion via slip causing all atoms above and below the slip plane to be permanently displaced from one set of lattice sites to a new set of lattice sites. Thus to explore the lattice displacement in a microscopic view would provide deep insight into understanding the dislocation movement. Consider an edge dislocation moving one step in a crystal as shown in Fig. 5.10. The dark circles of Fig. 5.10 show the location of atoms when the extra half-plane is located at position 1 moving to the position 2 with open circles. Taking into account the atoms at the slip plane, the atoms at the extra half-plane (position 1) would push its first nearest neighbors, 2L and 1R, off their lattice sites by 1/3 a (a is the lattice parameter), the second nearest neighbors, 3L and 2R, off by 1/5 a, and the third neighbors, 4L and 3R, are unaffected. As observed in Fig. 5.10, the extra plane at position 1 apparently moves one step to position 2 after the small shift of the atoms around the

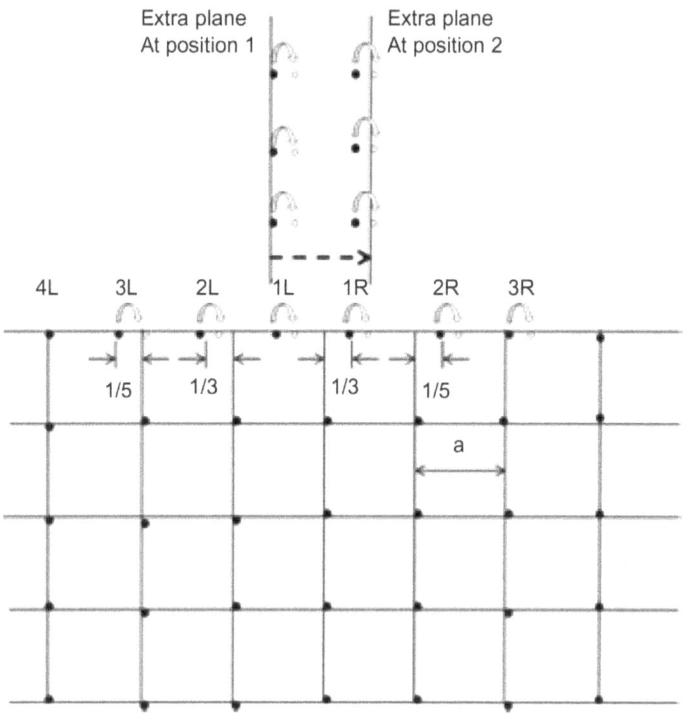

FIGURE 5.10 Schematic illustration of the local displacement of atoms associated with an edge dislocation movement. *From Verhoeven, J.D., 1975. Fundamentals of Physical Metallurgy, John Wiley & Sons, New York.*

dislocation. This is the reason why causing plastic flow through the dislocation motion in crystals required lower energy.

This shift is accomplished by a local disturbance, restricted to the region immediately around the dislocation, where there are several examples of motion in nature analogous to dislocation slip motion, like the motion of a caterpillar, the motion of a worm, or the motion of a rug.

Upon understanding the dislocation slip motion, we can further correlate the dislocation slip motion with the macroscopic crystal deformation, in reference to the relative displacement of top and bottom surface of the crystal. It has been deduced (Cottrell, 1953; Hull and Bacon, 1984) that if the dislocation moves through a distance Δx (Fig. 5.11B(I)), the crystal will be sheared by an amount equal to $b(\Delta x/x)$ (Fig. 5.11B(II)), where x is the total distance across the slip plane. Nevertheless, the related mechanics is ambiguous, in particular, the physical meaning of b is often falsely considered as the Burgers vector occurring on the crystal surface at the very beginning, shown in Fig. 5.11A.

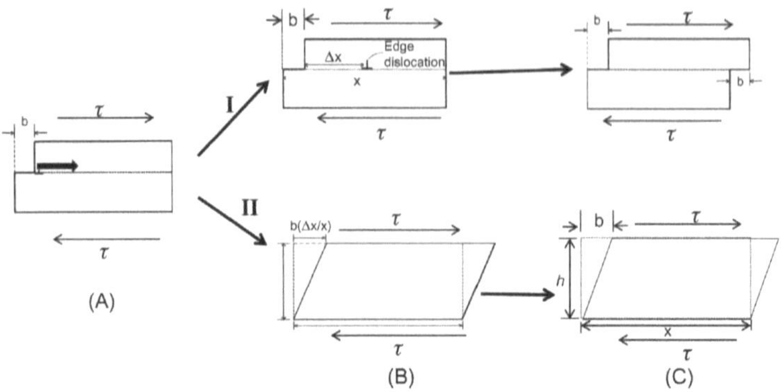

FIGURE 5.11 Correlation of the dislocation slip motion with the macroscopic crystal deformation: (A) a dislocation starts to move; (B) an edge dislocation moves Δx (I), and shear displacement of the crystal is $b(\Delta x/x)$ (II); and (C) the dislocation sweeps through the crystal, producing a unit step of slip, b, on the right-hand surface (I), and shear displacement of the crystal is b (II).

Hence, deep insight into understanding the correlation of dislocation motion and crystal deformation will be provided here. Regarding the dislocation slip motion, there are two displacements needed to be distinguished as the dislocation moves one step: one is along the direction of Burgers vector, called the lattice displacement, referred to as the relative displacement in the local regions around the dislocation above and below the slip plane, basically the local Burgers vector of b. The other is along the moving direction of the dislocation line, being in a highly deformed region that is hard to be evaluated and thus assumed to be ds. For a solid, it can be considered that interactions between the atoms are assumed to act like being linked with linear springs in a three-dimensional network.[1] On the basis of this consideration, the interrelationship of the response of the crystal to the applied forces and the movement of dislocations can be perceived. Intuitively, it is suggested that the crystal deformation stems from the lattice displacement. Thus it is necessary to pursue how the local disturbance due to lattice displacement along the direction of Burgers vector induces the crystal deformation. It is inferred that the induced shear displacement, d, of the crystal as the dislocation moves one step can be described in a form of $bf(\delta)$, where $f(\delta)$ is a polynomial function of δ, and δ is an unknown

1. Basically, this may conform to the Born–von Kármán model assuming nuclear masses connected by massless electron springs, i.e., the forces between nuclei are transmitted through lightweight electrons, which are further assumed to respond instantly to nuclear displacements (Born–Oppenheimer approximation). As the atom at the slip plane moves, whose electronic structure will be redistributed that in turn creates a force to exert on the nearest-neighbor atom to move. This action would be kept on to the surface atom. The Born–von Kármán model is expected to be reliable at low temperatures where atom displacements are small.

variable,[2] depending on bond strength and the crystal structure. Since d is much smaller than b, $f(\delta)$ is minute. Thus to a first-order approximation, the shear displacement of the crystal deformation can simply be represented by the lattice displacement multiplied by the variable δ, namely, $b\delta$. Moreover, in order to preserve the magnitude of the local Burgers vector being constant, it is necessary for the lattice displacement to maintain the same value as the dislocation moves every step. Thus the value of $b\delta$ is constant for each step. Furthermore, the deformation of shear can essentially be considered as the accumulation of the induced displacement of every single step of the dislocation movement. The corresponding total shear displacement, D_t, would be $D_t = \sum_i^n d_i = nb\delta$, where $n = x/ds$. In addition, as the dislocation sweeps through the crystal, a unit step of slip, b, is also produced on the right-handed surface (Fig. 5.11C(II)); therefore, it can be inferred that D_t is equivalent to b, namely, $D_t = b = \frac{x}{ds} b\delta$, giving $\delta = \frac{ds}{x}$. As a result, the shear displacement of the crystal deformation as the dislocation moves one step is $b\frac{ds}{x}$. This can be extended to that as the dislocation moves Δx, the induced shear displacement of the crystal is $D = b\frac{\Delta x}{x}$, represented in Fig. 5.11B(II), which equivalently can be expressed as $D = b\frac{\Delta A}{A}$, where A is the area of the slip plane and ΔA is the fraction of area passed over by the dislocation. Since the shear strain $\Delta\gamma$ given to the crystal equals the displacement D divided by the height h of the crystal, we have

$$\Delta\gamma = \frac{b\Delta A}{Ah} = \frac{b\Delta A}{V}, \tag{5.7}$$

where V ($=Ah$) is the volume of the crystal.

It is noted that a measurable strain can be observed only when a sufficiently large number of dislocations move. Assume that n edge dislocations of length l move through an average distance $\Delta\bar{x}$, this relation becomes

$$\Delta\gamma = \frac{bnl\Delta\bar{x}}{V} = \rho b\Delta\bar{x}, \tag{5.8}$$

where $\rho = nl/V$ is the density of the mobile dislocations. Note that the dislocation density can be defined as: Suppose n dislocation lines intersect a surface of area A and are perpendicular to the surface; then consider an element of crystal of L deep. The dislocation density is expressed as

$$\rho = \frac{\text{length}}{\text{vol}} = \frac{nL}{AL} = \frac{n}{A} \tag{5.9}$$

2. At the very beginning, a dislocation moves under a virtue force, F, (see Section 5.10). Assuming the induced shear displacement, d, results from the force (N) responding to the atom displacement (footnote 1), the function of d and N basically can be fitted by a polynomial function, i.e., $d = g(N) = \alpha N + \beta N^2 + \gamma N^3 + \ldots$ Since a dislocation moves one step b under F, $b = g(F)$. Further, $d/b = g(N)/g(F) = \alpha N/g(F) + \beta N^2/g(F) + \gamma N^3/g(F) + \ldots = \alpha N/g(F) + (\beta g(F)/\alpha^2)(\alpha N/g(F))^2 + (\gamma(g(F))^2/\alpha^3)(\alpha N/g(F))^3 + \ldots$ Thus d can be described in a form of $bf(\delta)$, where $f(\delta) = \alpha N/g(F) + (\beta g(F)/\alpha^2)(\alpha N/g(F))^2 + (\gamma(g(F))^2/\alpha^3)(\alpha N/g(F))^3 + \ldots$ is a polynomial function of $\delta = \alpha N/g(F)$.

Eq. (5.9) indicates that dislocation density can be defined as the number of dislocations per unit area (cm^{-2}). If the edge dislocations move through an average distance $\Delta\bar{x}$ in a time interval Δt, we have

$$\frac{\Delta\gamma}{\Delta t} = \dot{\gamma} = \rho b\bar{v}, \tag{5.10}$$

where $\dot{\gamma}$ is the shear strain rate and $\bar{v}$ is the average dislocation velocity. For the tensile strain rate in a polycrystalline metal, it is reasonably appropriated as

$$\dot{\varepsilon} = \frac{1}{2}\dot{\gamma} = \frac{1}{2}\rho b\bar{v}, \tag{5.11}$$

where the factor 1/2 is Schmid orientation factor.

5.7 STRESS FIELDS OF DISLOCATIONS

5.7.1 Constitutive Equations With Lame Constants

For structure analysis, the constitutive relations are referred to the relations between stress and strain under more general conditions where the stresses and strains may vary with position and time and temperature. For the solution of many problems in continuum mechanics, the generalized relation between stress and strain is needed. By assuming that the material is isotropic, the constitutive relations can be completely characterized by only two elastic constants, λ and μ, known as the Lame constants. The complete set of constitutive equations is expressed as

$$\begin{aligned}
\sigma_x &= (\lambda + 2\mu)\varepsilon_x + \lambda\varepsilon_y + \lambda\varepsilon_z \\
\sigma_y &= \lambda\varepsilon_x + (\lambda + 2\mu)\varepsilon_y + \lambda\varepsilon_z \\
\sigma_z &= \lambda\varepsilon_x + \lambda\varepsilon_y + (\lambda + 2\mu)\varepsilon_z \\
\tau_{yz} &= \mu\gamma_{yz} \\
\tau_{zx} &= \mu\gamma_{zx} \\
\tau_{xy} &= \mu\gamma_{xy}
\end{aligned} \tag{5.12}$$

λ and μ (usually known as the shear modulus) are related to Young's modulus, E, and Poisson's ratio, υ as follows.

$$\mu = \frac{E}{2(1+\upsilon)} \tag{5.13}$$

$$\lambda = \frac{\upsilon E}{(1+\upsilon)(1-2\upsilon)} \tag{5.14}$$

Concerning the subscripts of shear stress and shear strain, the first subscript on the shear components specifies the direction of the normal to the plane on which the shear stress exerts; the second subscript denotes the direction of the shear stress. Note that in the tensor analysis, it is more common to take the opposite way.

In order to make an analysis of elastic stresses and strains applicable to dislocations in crystalline materials, dislocations can be viewed macroscopically as

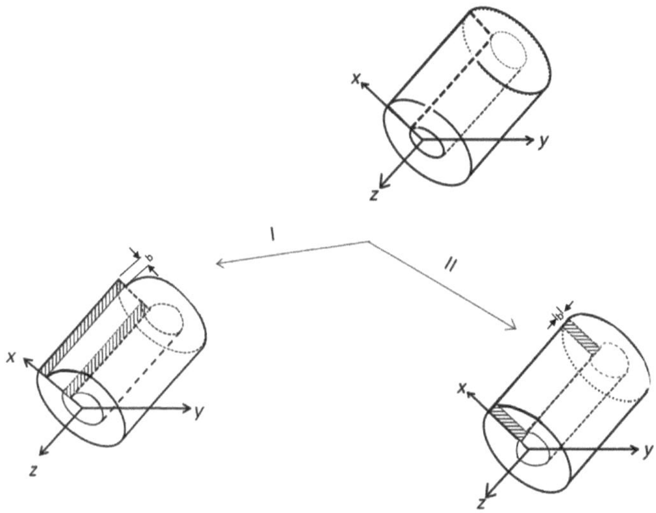

FIGURE 5.12 Schematic illustration of the nature of Volterra dislocations in an isotropic continuum: simulation of edge (I) and screw (II) dislocation displacements.

"disturbances" in an otherwise continuous, isotropic medium. In 1907, Volterra employed a stress-free cylindrical volume with an axial hole avoiding the occurrence of mathematical complications in the course of the analysis. This cylinder is cut as shown in Fig. 5.12 (top). The two half planes are further displaced with respect to each other in two ways, as shown in Fig. 5.12(I) concerning the simulation of edge dislocation displacement and (II) the simulation of screw dislocation displacement. In the following, the stress fields of edge and screw dislocations will be deduced underlying this model.

5.7.2 Stress Field of a Screw Dislocation

In the analysis of the stress fields of the dislocations, continuum mechanics modeling will be utilized with the known values of the displacements u, v, and w in the x-, y-, and z-directions, respectively. As observed in Fig. 5.12(II), since the displacement of the produced screw dislocation at the cut surfaces is only in the z direction, we obtain the displacements for x-, y-, and z-direction:

$$u = 0, \qquad v = 0, \qquad w = \frac{b}{2\pi}\tan^{-1}\left(\frac{y}{x}\right) \tag{5.15}$$

The strains can be expressed from the displacements given by Eq. (5.15) by using the following differential equations:

For normal strains along x_i axis:

$$\varepsilon_{ii} = \frac{\partial u_i}{\partial x_i} \tag{5.16}$$

For the true shear strains, ε_{ij}, $i \neq j$:

$$\varepsilon_{ij} = \frac{1}{2}\left(\frac{\partial u_i}{\partial x_j} + \frac{\partial u_j}{\partial x_i}\right) = \varepsilon_{ji}$$

For the engineering shear strains γ_{ij}, $i \neq j$:

$$\gamma_{ij} = 2\varepsilon_{ij}$$

So

$$\gamma_{ij} = \frac{\partial u_i}{\partial x_j} + \frac{\partial u_j}{\partial x_i} = \gamma_{ji} \tag{5.17}$$

Based on Eqs. (5.16) and (5.17), the strains associated with screw dislocation are readily evaluated and have the values

$$\begin{aligned} \varepsilon_x = \varepsilon_y = \varepsilon_z = \gamma_{xy} &= 0 \\ \gamma_{zx} &= -\frac{b}{2\pi}\left(\frac{y}{x^2+y^2}\right) \\ \gamma_{yz} &= \frac{b}{2\pi}\left(\frac{x}{x^2+y^2}\right) \end{aligned} \tag{5.18}$$

The stresses corresponding to the strains represented in Eq. (5.18) are given directly by using the constitutive equations indicated in Eq. (5.12):

$$\begin{aligned} \sigma_x = \sigma_y = \sigma_z = \tau_{xy} &= 0 \\ \tau_{zx} &= -\frac{\mu b}{2\pi}\left(\frac{y}{x^2+y^2}\right) \\ \tau_{yz} &= \frac{\mu b}{2\pi}\left(\frac{x}{x^2+y^2}\right) \end{aligned} \tag{5.19}$$

The stresses of Eq. (5.19) can be converted from Cartesian to cylindrical coordinates as follows:

$$\begin{aligned} \sigma_r = \sigma_\theta = \sigma_z = \tau_{r\theta} = \tau_{zr} &= 0 \\ \tau_{\theta z} &= \frac{\mu b}{2\pi r} \end{aligned} \tag{5.20}$$

Through Eq. (5.20), we obtain an important result, i.e., the magnitude of $\tau_{\theta z}$ decreases as $1/r$ with increasing distance from the center of the screw dislocation, called a r^{-1} singularity. Following this result, the most basic dislocation formulas, namely, $\tau \sim \sqrt{\rho}$ (τ is shear stress; ρ is dislocation density), called the Taylor hardening relation, can be produced.

5.7.3 Stress Fields of an Edge Dislocation

As indicated in Fig. 5.12(I), no displacements occur in the z direction, i.e., $w = 0$. The correct expressions for u and v are listed in the following:

$$u = \frac{b}{2\pi}\left[\tan^{-1}\frac{y}{x} + \frac{xy}{2(1-\upsilon)(x^2+y^2)}\right]$$
$$v = -\frac{b}{2\pi}\left[\frac{1-2\upsilon}{4(1-\upsilon)}\ln(x^2+y^2) + \frac{x^2-y^2}{4(1-\upsilon)(x^2+y^2)}\right] \tag{5.21}$$

In the same procedure as for the screw dislocation, the strains around an edge dislocation can be deduced, and as a result, the following expressions for the stress field of an edge dislocation can be obtained:

$$\sigma_x = -\frac{\mu b}{2\pi(1-\upsilon)}\frac{y(3x^2+y^2)}{(x^2+y^2)^2}$$
$$\sigma_y = -\frac{\mu b}{2\pi(1-\upsilon)}\frac{y(x^2-y^2)}{(x^2+y^2)^2}$$
$$\sigma_z = \upsilon(\sigma_x+\sigma_y) = -\frac{\mu\upsilon b y}{\pi(1-\upsilon)(x^2+y^2)} \tag{5.22}$$
$$\tau_{xy} = \frac{\mu b}{2\pi(1-\upsilon)}\frac{x(x^2-y^2)}{(x^2+y^2)^2}$$
$$\tau_{yz} = \tau_{zx} = 0$$

The occurrence of a Poisson contraction, specified by υ, requires a stress σ_z, in the z-direction to maintain zero displacement in this direction.

5.8 ENERGY OF A DISLOCATION

Since the strains of a dislocation are restricted in its vicinity, the total increase in energy due to the presence of a dislocation can be divided into two parts: (1) the core of the dislocation (the region within about a radius $r_0 = 2|b| = 2b$ of the dislocation line) contribute about 10% of the energy; and (2) the elastic strains in the balance of the crystal (principally out to a radius of about $R = 10^{-3}$ to 10^{-6} cm, depending on the dislocation density) account for the remaining 90% of the self-energy of a dislocation in a typical polycrystalline metal. The self-energy of a dislocation can be estimated by the work involving force, F, and distance, s:

$$\text{work} = \int_0^s F ds = V\int_0^s \frac{F}{A}\frac{ds}{l} = V\int_0^\varepsilon \sigma d\varepsilon \tag{5.23}$$

where V is the volume of the crystal; A is its area; and l is its length. The strain energy density, E, is the work done by all nine stresses per unit volume:

$$E = \int_0^{\varepsilon_{ij}} \sum_{i,j=1}^{3} \sigma_{ij}\varepsilon_{ij} = \frac{1}{2}(\sigma_x\varepsilon_x + \sigma_y\varepsilon_y + \sigma_z\varepsilon_z + \tau_{xy}\gamma_{xy} + \tau_{yz}\gamma_{yz} + \tau_{zx}\gamma_{zx}) \quad (5.24)$$

where the strains are considered to depend linearly on the stresses, and the relationships of $\varepsilon_{ij} = \varepsilon_{ji}$ and $\gamma_{ij} = 2\varepsilon_{ij}$ are applied, and 1, 2, and 3 are replaced by x, y, and z, respectively.

5.8.1 Strain Energy of a Screw Dislocation

In Section 5.7.2, we have obtained the constitutive relations of the stresses and strains around a screw dislocation (Eqs. (5.18) and (5.19)), which indicates that only two terms are nonzero in Eq. (5.24). Thus by eliminating other terms, the self-energy of a screw dislocation per unit volume is given:

$$E = \frac{1}{2}(\tau_{yz}\gamma_{yz} + \tau_{zx}\gamma_{zx}) \quad (5.25)$$

Substitute Eqs. (5.18) and (5.19) into Eq. (5.25), yielding

$$E = \frac{\mu}{8}\left(\frac{b}{\pi}\right)^2\left(\frac{1}{x^2 + y^2}\right) \quad (5.26)$$

and in cylindrical coordinates,

$$E = \frac{\mu}{8}\left(\frac{b}{\pi}\right)^2\left(\frac{1}{r^2}\right) \quad (5.27)$$

The total elastic strain energy in a cylinder shown in Fig. 5.13 can be obtained by integrating over the volume of the cylinder:

$$E_v = \int_v E dV = \frac{\mu}{8}\left(\frac{b}{\pi}\right)^2 \int_{r_0}^{R} \frac{2\pi r}{r^2} l dr = \frac{\mu b^2 l}{4\pi} \ln\frac{R}{r_0} \quad (5.28)$$

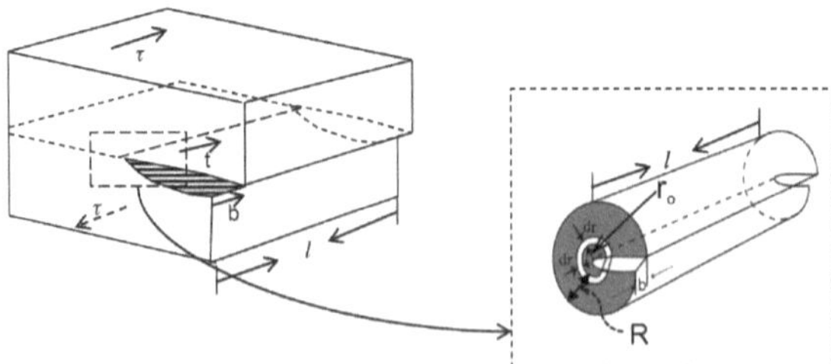

FIGURE 5.13 Schematic illustration of modeling of the elastic distortion associated with a pure screw dislocation.

In order to reflect the fact that dislocations are line defects, the energy is to be represented as per unit length:

$$E_s = \frac{E_v}{l} = \frac{\mu b^2}{4\pi} \ln \frac{R}{r_0} \tag{5.29}$$

Taking r_0 to be about 10^{-8} cm and R to be 10^{-3} to 10^{-6} cm, depending on the dislocation density, we obtain

$$E_s \simeq \frac{\mu b^2}{2} \quad \text{to} \quad \mu b^2 \tag{5.30}$$

As implied in Eq. (5.29), since E_s diverges as R approaches ∞ or as r_0 approaches 0, the divergence with R shows that one cannot ascribe to the dislocation a definite characteristic energy; the energy depends on the size of the crystal. For the case of one dislocation in a crystal, an approximate choice for R is the shortest distance from the dislocation to the free surface. In a crystal containing many dislocations of both signs, a reasonable approximation would be to take roughly half the average distance between dislocations for R, where the stress field of the dislocation is assumed to be neutralized. Since this distance is equal to half the spacing of adjacent dislocations, the value of R and therefore of E_s increases as the density of dislocations decreases. Note that the mean spacing, l, between dislocations can be defined as

$$l = \frac{1}{\sqrt{\rho}} \tag{5.31}$$

5.8.2 Strain Energy of an Edge Dislocation

Following the same procedure as did for the screw dislocation, the self-energy of an edge dislocation per unit length, E_E, can be evaluated. The result is found out as:

$$E_E = E_s/(1-\upsilon) = \frac{\mu b^2}{4\pi(1-\upsilon)} \ln \frac{R}{r_0} \tag{5.32}$$

For a typical metal, Poisson's ratio is

$$\upsilon \cong \frac{1}{3} \qquad E_E \cong (3/2)E_s \tag{5.33}$$

5.9 LINE TENSION OF A DISLOCATION

In analogy to the interrelationships of surface tension and surface free energy discussed in Section 6.3 of chapter 6, the energy per unit length (E, J/m) of a dislocation line corresponds to a force (N', newton = J/m) tending to shorten the length of the line. The force, N', is directed along the dislocation line and is called the line tension. When shortening occurs, the self-energy, associated with an a dislocation, would be released, which infers that

$$N' = -\frac{\Delta E_v}{\Delta l} = -\frac{\mu b^2}{4\pi}\ln\left(\frac{R}{r_0}\right) \tag{5.34}$$

With the aforementioned estimates of R and r_0, the approximate values of N' depending on the dislocation density for both screw and edge dislocation are

$$N' \cong -\frac{\mu b^2}{2} \quad \text{to} \quad -\mu b^2 \tag{5.35}$$

5.10 FORCES UPON DISLOCATIONS

Forces may exert on a dislocation for various reasons. Externally applied stresses, various sources of internal stresses, free surfaces, and internal discontinuities all provide opportunities for the energy to be altered by the movement of dislocations.

When a shear stress is applied, a virtual force (F_v) develops and is exerted on the dislocation because as the dislocation moves (establishes kinetic energy), potential energy, which provides the stress, decreases.

Thus, the work done by an applied stress in a slip plane, when a dislocation moves in the plane, is partly used to increase the kinetic energy of the dislocation and is partly for the energy dissipation due to the refraining force consisting of line tension and frictional force including impurity interaction, Peierls-Nabarro stress, and dissipation in lattice vibrations, whose processes may involve thermoelastic damping, radiation damping, and scattering of

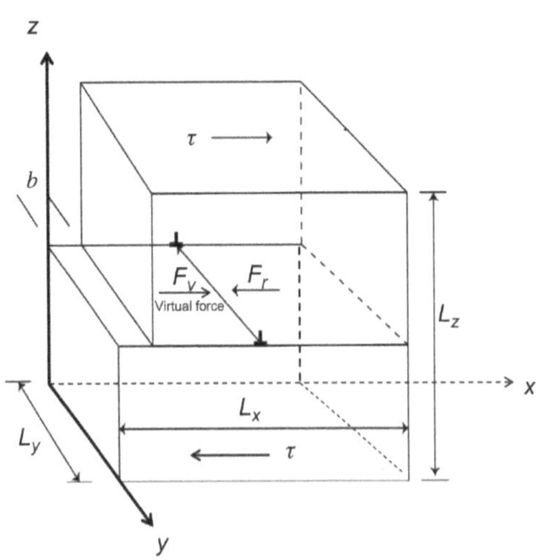

FIGURE 5.14 Schematic presentation of the presence of a refraining force (F_r) needed to maintain equilibrium with a virtual force (F_v) as a shear stress exerts on the crystal.

sound waves. This implies that there is a refraining force (F_r in Fig. 5.14) needed to maintain equilibrium with virtual force, namely, at equilibrium, $F_v = -F_r$. Once F_v overcomes F_r, dislocations are allowed to move. As discussed in Section 5.6, if a dislocation moves only through a distance Δx, then the average shear displacement of the crystal due to glide of dislocation line is equal to $b(L_y\Delta x/L_xL_y)$ or $b(\Delta x/L_x)$. Thus when the dislocation moves over distance Δx, work done by the shear stress is $\tau w L_x\left(b\cdot\frac{\Delta x}{L_x}\right)$, which would be removed by the refraining force doing the same work, that is

$$F_r\Delta x = \tau L_y L_x\left(b\cdot\frac{\Delta x}{L_x}\right) \tag{5.36}$$

$$\frac{F_r}{L_y} = \tau b = \text{force per unit length} \tag{5.37}$$

5.11 THE BOWING OF A DISLOCATION

Assuming that a virtual force, F, is exerted on a dislocation, the dislocation line is pinned and bowed. In such a case, the frictional forces can be neglected.

As the dislocation line is bowed, a force N tending to straighten the line must overcome the force produced by line tension in order to maintain a given curvature in the dislocation line, i.e., $N = -N'$. In taking into account the corresponding balance-of-force diagram, shown in Fig. 5.15, we obtain:

$$2N\sin\frac{\phi}{2} = Fr\phi \tag{5.38}$$

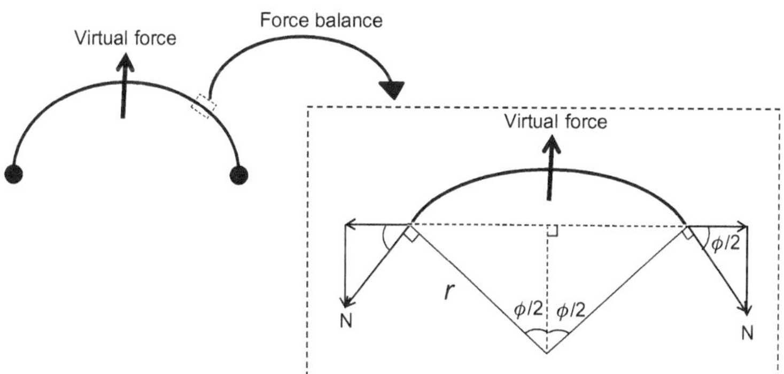

FIGURE 5.15 Sketch of the analysis of the virtual force, F, per unit length required to maintain a radius of curvature, r, in a very small segment of a pinned dislocation line and the corresponding diagram of balance of force.

In Fig. 5.15, a very small segment is taken from the pinned dislocation line, and thus ϕ is very small, leading to

$$\sin\frac{\phi}{2} \cong \frac{\phi}{2} \tag{5.39}$$

Therefore, we obtain

$$N\phi = Fr\phi \tag{5.40}$$

$$F = \frac{N}{r} \tag{5.41}$$

For a crystal with high dislocation density,

$$N = \mu b^2 \tag{5.42}$$

$$F = \frac{\mu b^2}{r} \tag{5.43}$$

Furthermore, because

$$F = \tau b \tag{5.44}$$

$$\tau = \frac{\mu b}{r} \tag{5.45}$$

The pinning of dislocations either by other dislocations or solute atoms in materials is an important subject of the strengthening of materials and Frank–Read dislocation sources.

5.12 FORCE BETWEEN DISLOCATIONS

The interactions between dislocations are complicated and the general treatment of the forces involved is rather abstract. Thus it is helpful to have a useful physical picture, in which the forces can be visualized in terms of energies. At first, we consider two edge dislocations on the same slip plane with the Burgers vectors being equal but opposite. It is assumed that these two edge dislocations are separated initially and then approach each other to combine and annihilate. As per Eq. (5.32), the energy difference between initial separation and final annihilation is obtained as follows.

$$\Delta E = 0 - 2\frac{\mu b^2}{4\pi(1-\upsilon)}\ln\frac{R}{r_0} = -\frac{\mu b^2}{2\pi(1-\upsilon)}\ln\frac{R}{r_0} \tag{5.46}$$

Since the energy of the system varied with r, we can picture it as a potential energy (P.E.) function of r, in which r is in a range from r_0 to the normal half-distance, R, between dislocations. Thus

$$\text{P.E.} = \frac{\mu b^2}{2\pi(1-\upsilon)}\ln\frac{r}{r_0} \tag{5.47}$$

We can further obtain the force:

$$F' = -\frac{d(\text{P.E.})}{dr} = -\frac{d}{dr}\left[\frac{\mu b^2}{2\pi(1-\upsilon)}\ln\frac{r}{r_0}\right] = -\frac{\mu b^2}{2\pi(1-\upsilon)r} \tag{5.48}$$

The negative sign indicates that the force of interaction is attractive.

For the second case, the Burgers vectors of the two edge dislocations of the same sign are considered. In this manner, the energy increases if two dislocations combine. The Burgers vector is $2\vec{b}$, and the energy is increased by the factor $(2b)^2$. In the same way as in the first case, the force of interaction can be obtained as follows:

$$\Delta E = \frac{\mu(2b)^2}{4\pi(1-\upsilon)}\ln\frac{R}{r_0} - 2\frac{\mu b^2}{4\pi(1-\upsilon)}\ln\frac{R}{r_0} = \frac{\mu b^2}{2\pi(1-\upsilon)}\ln\frac{R}{r_0} \tag{5.49}$$

$$\text{P.E.} = -\frac{\mu b^2}{2\pi(1-\upsilon)}\ln\frac{r}{r_0} \tag{5.50}$$

$$F' = \frac{\mu b^2}{2\pi(1-\upsilon)r} \tag{5.51}$$

The positive sign suggests repulsion in this case.

5.13 PEACH–KOEHLER EQUATION

As aforementioned in Section 5.5, the shear stress components in the slip direction of b, called the resolved shear stress (τ), would exert a virtual force F on the edge dislocation line normally, in which we obtain a relationship of $F = \tau b$, underlying the energy viewpoint. However, the applied stress may be in a tensor form, and the force between two dislocations depends not only on their Burgers vector but also on the sense vector $\vec{t}$ of each dislocation line and on the orientation of the two lines in three dimensions. The general form represented in vector notation is known as the Peach–Koehler equation, which can further be derived by the same procedure for deriving the relationship between virtual force and the applying stress (Section 5.10) in terms of vector notations. Assuming a dislocation of line length L moves by $d\vec{r}$, the work done by the virtual force per unit length is $\frac{\vec{F'}}{L}\cdot d\vec{r}$. The force caused by the stress tensor Σ on the area $(\vec{t}\times d\vec{r})$ per unit length is $\Sigma\cdot(\vec{t}\times d\vec{r})$. Supposing that the crystal deforms by a displacement of $\vec{b}$, the balance of the work done by external shear stress and the force on the dislocation is

$$\left[\Sigma\cdot(\vec{t}\times d\vec{r})\right]\cdot\vec{b} = \frac{\vec{F'}}{L}\cdot d\vec{r} \tag{5.52}$$

$$\frac{\vec{F'}}{L}\cdot d\vec{r} = (\vec{b}\cdot\Sigma)\cdot(\vec{t}\times d\vec{r}) = \left[(\vec{b}\cdot\Sigma)\times\vec{t}\right]\cdot d\vec{r} \tag{5.53}$$

$$\vec{F} = \frac{\vec{F}'}{L} = (\vec{b} \cdot \Sigma) \times \vec{t} \tag{5.54}$$

where $\vec{F}$ is the force per unit length at an arbitrary point along the dislocation line, Σ is the applied stress tensor, and $\vec{t}$ is the unit vector of the local dislocation line tangent direction at a given point. In practice, Eq. (5.54) should be expressed in matrix forms:

$$\vec{b} = \begin{bmatrix} b_x & b_y & b_z \end{bmatrix}$$

$$\Sigma = \begin{bmatrix} \sigma_x & \tau_{xy} & \tau_{xz} \\ \tau_{yx} & \sigma_y & \tau_{yz} \\ \tau_{zx} & \tau_{zy} & \sigma_z \end{bmatrix}$$

$$G = \vec{b} \cdot \Sigma = \begin{bmatrix} b_x & b_y & b_z \end{bmatrix} \begin{bmatrix} \sigma_x & \tau_{xy} & \tau_{xz} \\ \tau_{yx} & \sigma_y & \tau_{yz} \\ \tau_{zx} & \tau_{zy} & \sigma_z \end{bmatrix}$$

$$\begin{aligned} G_x &= \sigma_x b_x + \tau_{xy} b_y + \tau_{xz} b_z \\ G_y &= \tau_{yx} b_x + \sigma_y b_y + \tau_{yz} b_z \\ G_z &= \tau_{zx} b_x + \tau_{zy} b_y + \sigma_z b_z \end{aligned} \tag{5.55}$$

$$\begin{aligned} \vec{F} = G \times \vec{t} &= \begin{vmatrix} \vec{i} & \vec{j} & \vec{k} \\ G_x & G_y & G_z \\ t_x & t_y & t_z \end{vmatrix} \\ &= (G_y t_z - G_z t_y)\vec{i} + (G_z t_x - G_x t_z)\vec{j} + (G_x t_y - G_y t_x)\vec{k} \\ &= -\vec{t} \times G \end{aligned}$$

Here, $\vec{i}$, $\vec{j}$, and $\vec{k}$ are the unit vectors along the *x*-, *y*-, and *z*-directions; and G is the local force per unit length exerting on a plane normal to the Burgers vector. The vector cross product emphasized that the virtual force acts in the direction perpendicular to the direction of the dislocation line.

As per Eq. (5.55), and according to the established coordinate axes, shown in Fig. 5.14, three cases with one component of the stress tensor can easily be obtained as follows:

Case one: for an edge dislocation with a shear stress τ_{zx}, Burgers vector $[b_x00]$, and line direction [010]:

$$F_x = -\tau_{zx} b_x \tag{5.56}$$

Case two: for an edge dislocation with a normal stress σ_x, Burgers vector $[b_x00]$, and line direction [010]:

$$F_z = \sigma_x b_x \text{ (climb)} \tag{5.57}$$

Case three: for a screw dislocation with a shear stress τ_{zy}, Burgers vector $[0b_y0]$ and line direction [010]:

$$F_x = -\tau_{zy} b_y \tag{5.58}$$

The Peach–Koehler relation also provides the basis for understanding the interactions between dislocations. It can be considered that the forces exerting on the second dislocation are given via the Peach–Koehler formula in which the stress tensor Σ is provided by the first dislocation. In the following, we will give two cases concerning interactions between two parallel dislocations.

1. Two parallel screw dislocations with both Burgers' vectors and the dislocation lines in the z direction: $\vec{b_1} = [00b_1]$ and $\vec{b_2} = [00b_2]$, and $\vec{t} = [001]$

Take the stress field produced by the first dislocation as the stress tensor exerting on the second dislocation, where only two independent stress components are present and the others are zero. These two independent components are

$$\begin{aligned} \tau_{zx} &= -\frac{\mu b_1}{2\pi}\left(\frac{y}{x^2+y^2}\right) \\ \tau_{yz} &= +\frac{\mu b_1}{2\pi}\left(\frac{x}{x^2+y^2}\right) \end{aligned} \tag{5.59}$$

Furthermore, utilize the Peach–Koehler equation to calculate the virtual force on the second dislocation, and we obtain

$$\vec{F} = \tau_{yz} b_2 \vec{i} - \tau_{zx} b_2 \vec{j} \tag{5.60}$$

Apply Eqs. (5.59) and (5.60), yielding

$$\vec{F} = \frac{\mu b_1 b_2}{2\pi(x^2+y^2)}\left(x\vec{i} + y\vec{j}\right) \tag{5.61}$$

Convert Eq. (5.61) into a cylindrical form, giving

$$\vec{F} = \frac{\mu b_1 b_2}{2\pi r}\vec{r} \tag{5.62}$$

If the Burgers vectors are of the same sign, F is positive representing a repulsive force; by contrast, for the opposite sign, F is negative and the two dislocations experience an attraction.

2. Two parallel edge dislocations with the Burgers' vectors in the x-direction but the dislocation lines in the z-direction: $\vec{b_1} = [b_100]$ and $\vec{b_2} = [b_200]$, and $\vec{t} = [001]$

Using the same procedure as (1), we can obtain the virtual force for the interactions between two parallel edge dislocations.

$$\vec{F} = \frac{\mu b_1 b_2}{2\pi(1-\upsilon)(x^2+y^2)^2}\left[x(x^2-y^2)\vec{i} + y(3x^2+y^2)\vec{j}\right] \tag{5.63}$$

Converting Eq. (5.63) into a cylindrical form, gives

$$\begin{aligned} F_x &= \frac{\mu b_1 b_2}{2\pi(1-\upsilon)r}(\cos^2\theta - \sin^2\theta)\cos\theta \\ F_y &= \frac{\mu b_1 b_2}{2\pi(1-\upsilon)r}\sin\theta(1+2\cos^2\theta) \end{aligned} \tag{5.64}$$

$$\begin{aligned} &At\ \theta = 0^\circ \ \text{ or } \ 180^\circ \\ &F_x = \frac{\mu b_1 b_2}{2\pi(1-\upsilon)r} \\ &F_y = 0 \end{aligned} \tag{5.65}$$

This situation concerns these two edge dislocations on the same slip plane, wherein the obtained force in the x-direction is consistent with that derived in Section 5.12 from the viewpoint of the energy.

$$\begin{aligned} &At\ \theta = 90^\circ \ \text{ or } \ 270^\circ \\ &F_x = 0 \\ &F_y = \frac{\mu b_1 b_2}{2\pi(1-\upsilon)r} \end{aligned} \tag{5.66}$$

The obtained force components tend to align two edge dislocations vertically, implying the occurrence of aggregates of parallel dislocations in such a configuration as the low-angle boundaries (Section 6.1.1 of chapter 6).

5.14 REACTIONS BETWEEN DISLOCATIONS

5.14.1 Dislocation Nodes

As aforementioned, the dislocation is the boundary between slipped and unslipped regions on the slip plane. Therefore, dislocations are supposed not to end within a crystal but they can end at the nodes developed with other dislocations. A dislocation node essentially is the junction point where multiple dislocations meet together, and is the essential portion of where a dislocation reaction occurs. As observed in Fig. 5.16A, a reaction occurs at the node, where two dislocations react/combine to form another:

$$\vec{b_1} + \vec{b_2} = \vec{b_3} \tag{5.67}$$

Whether or not this reaction is energetically feasible, it is determined by the energies of dislocations. Without other influences like external forces, thermal energy, etc., a requirement for the occurrence of this reaction is a net decrease in the self-energies, i.e.,

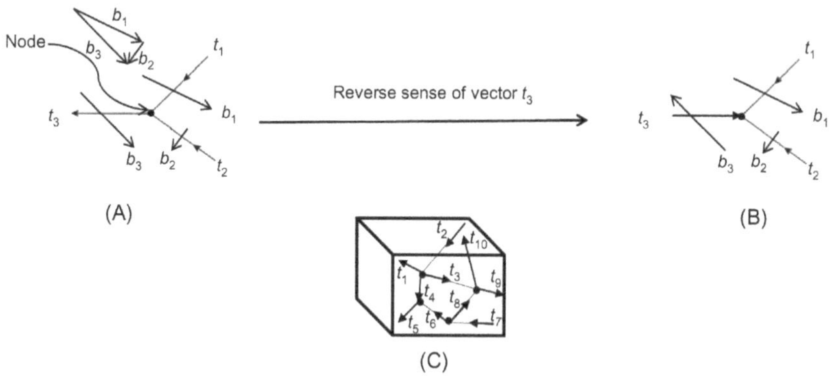

FIGURE 5.16 Schematic presentation of (A) the dislocation reaction at a node, where two dislocations can combine to form another, (B) reversion of the sense of vector t_3 in (A), where the Burgers vectors become $\vec{b_1} + \vec{b_2} + \vec{b_3} = 0$, and (C) schematic presentation of a network of dislocation lines within a crystal.

$$\mu b_1^2 + \mu b_2^2 > \mu b_3^2 \tag{5.68}$$

or

$$b_1^2 + b_2^2 > b_3^2 \tag{5.69}$$

In addition, if the vector sense of t_3 is reversed, its Burgers vector would be reversed, shown in Fig. 5.16B, leading to

$$\vec{b_1} + \vec{b_2} + \vec{b_3} = 0 \tag{5.70}$$

Eq. (5.70) indicates that the sum of the Burgers vectors running to the same node is zero, which is the conservation law of the Burgers vectors at the dislocation nodes, known as Frank's rule, with respect to the relationship between the Burgers vectors of multiple dislocations meeting at the same node, where the net Burgers vector must be zero, i.e.,

$$\vec{b}_{\text{net}} = \sum_{i=1}^{n} \vec{b}_i = 0 \tag{5.71}$$

As per Eq. (5.71), one unknown Burgers vector can be determined as all the other Burgers vectors are given. However, for the complex networks shown in Fig. 5.16C, the Frank's rule may alternatively be described as being the sum of the Burgers vectors of the entering dislocations is equal to those of the leaving dislocations if their dislocation lines are assumed to run from or to the node, i.e.,

$$\left(\sum_{i=1}^{n} \vec{b}_i\right)_{\text{entering}} = \left(\sum_{j=1}^{n} \vec{b}_j\right)_{\text{leaving}} \tag{5.72}$$

The direction of each dislocation line determines whether that dislocation is entering or leaving the node in question. Thus the following reactions would be held:

$$\begin{aligned} \vec{b_7} &= \vec{b_6} + \vec{b_8} \\ \vec{b_3} + \vec{b_8} &= \vec{b_9} + \vec{b_{10}} \end{aligned} \tag{5.73}$$

Here $\vec{b_1}$ represents the Burgers vector of the dislocation line $\vec{t_1}$, etc. Moreover, nodes have varying degrees of stability. The node is unstable and disappears when the reaction has been completed. Stable nodes are produced under many conditions, e.g., in the reaction of nonparallel dislocations or of dislocations on different slip planes, where the resultant dislocation is immobile. Since dislocations forming stable nodes would require substantially more force to move, the metals would be strengthened.

5.14.2 Kinks and Jogs

At the atomic level, dislocations are not straight. They contain steps of atomic dimensions. Two different types of short segments, kinks and jogs, shown in Figs. 5.17 and 5.18, respectively, can connect one straight portion of a dislocation line with a second portion offset by a small amount.

If the offset portion lies in the original slip, then the connecting segment is called a kink (Fig. 5.17A). After a pair of kinks has been produced, they can spread laterally (Fig. 5.17B) or can approach each other and be annihilated (Fig. 5.17C). As a dislocation moves (glides) on its slip plane, usually the dislocation does not remain a perfectly straight line but instead develops

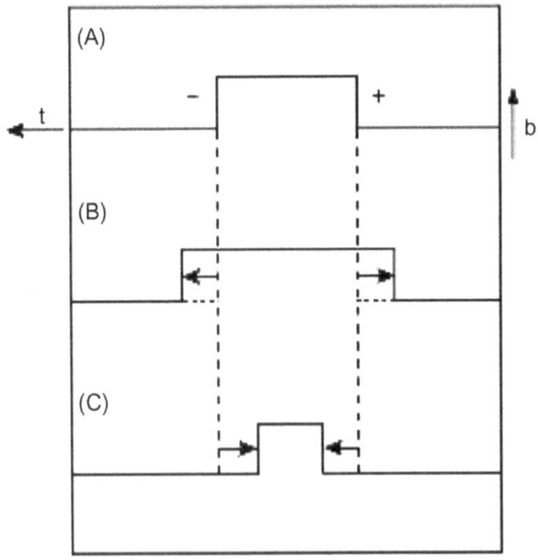

FIGURE 5.17 Schematic illustrations of (A) an edge dislocation with a ± pair of kinks with screw character lying in the slip plane, and the ± pair of kinks are (B) spreading and (C) approaching as the dislocation moves.

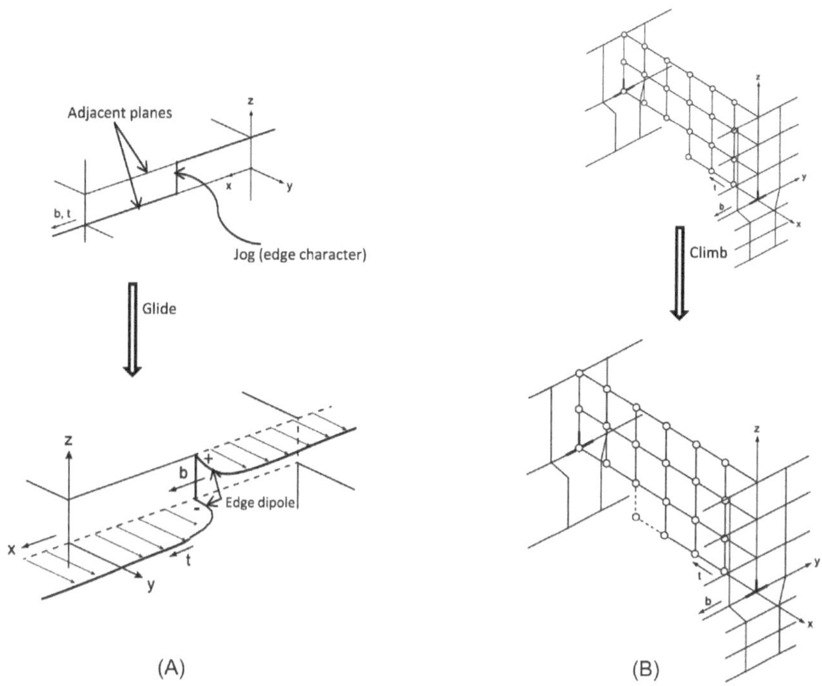

FIGURE 5.18 Schematic illustrations of jogs normal to the slip plane of the dislocations with different characteristics. (A) A screw dislocation containing the immobile jog results in the formation of an edge dipole as the main part of the dislocation glides. (B) An edge dislocation containing the immobile jog moves via climb.

some degree of curvature. Since curvature can be described by a suitable array of kinks, kinks are a common feature in the motion of dislocations. By contrast, the offset portion lies, not in the original slip plane, but in a parallel slip plane, then the connecting segment is called a jog (Fig. 5.18A). The jogs form initially as a result of: (1) intersection of two dislocations; or (2) climb of a portion of a straight dislocation line to produce an offset.

Regarding how a jog can impede the motion of a dislocation, it is illustrated in Fig. 5.18. As the screw dislocations move on their slip planes, each is pinned at the location of the jog. The result of this pinning action is the formation of an edge dipole, consisting of a pair of positive and negative edge dislocations. The screw dislocation can continue to glide, but only at the expense of creating an additional length of edge dipole (Fig. 5.18A). Alternatively, the original jog can move by climb (Fig. 5.18B).

5.14.3 Intersections of Dislocations

An important mechanism for forming kinks or jogs is the intersection of a moving dislocation line with another dislocation line in the crystal.

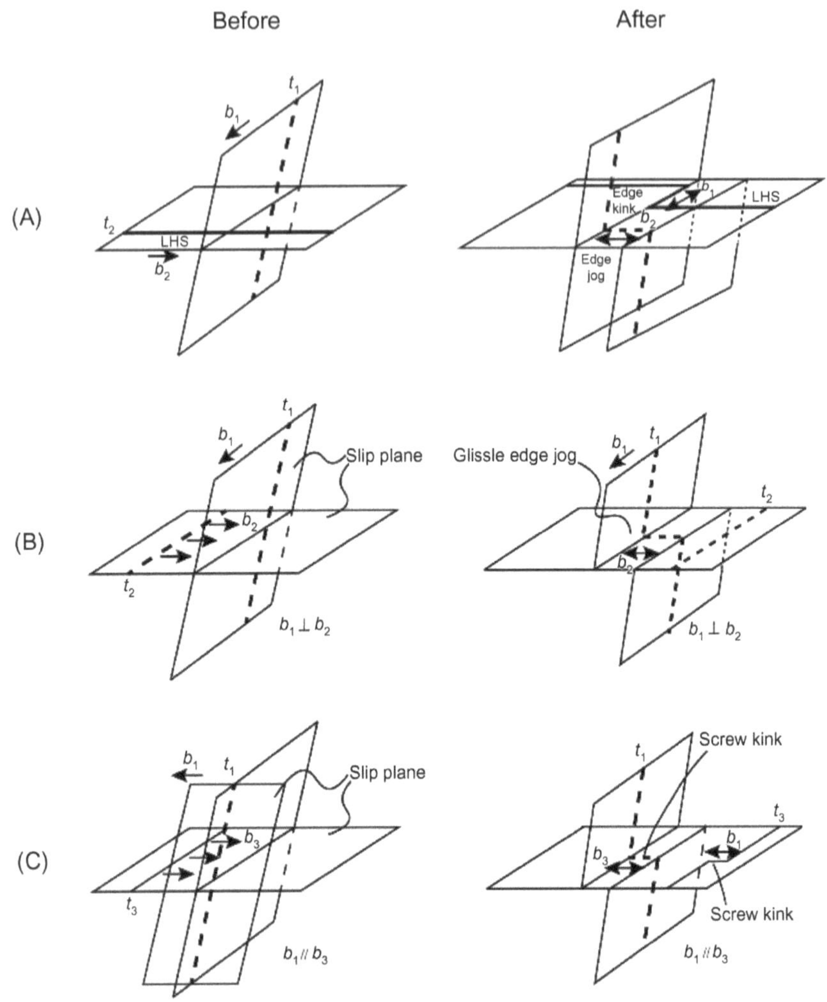

FIGURE 5.19 Schematic presentation of the creation of jogs and kinks by the intersection of (A) one edge and left-handed screw dislocations, (B) two edge dislocations with mutually perpendicular Burgers vectors and dislocation lines, and (C) two edge dislocations with mutually perpendicular dislocation lines but parallel Burgers vectors.

Because dislocations shear the lattice (and everything in it), they also shear the dislocations that they cut. Three cases will be presented in the following. Fig. 5.19A illustrates the intersection between an edge dislocation and a left-handed screw dislocation, where an edge jog is present at the former and an edge kink is present at the latter. Fig. 5.19B represents the intersection of both dislocation lines being edge character and mutually perpendicular. Since the two Burgers vectors are perpendicular, the motion of dislocation two causes an edge jog with magnitude of b_2 to appear in dislocation line 1 after intersection. (Dislocation line 2 is unaffected by the intersection because $\vec{b_1}$ is parallel to the direction of line 2.) Note that the resulting jog is glissile (i.e., able to glide) because it lies in a slip plane. In Fig. 5.19C, we take two parallel Burgers vectors into consideration. A kink of screw character appears in each of the dislocation lines after intersection.

If a sessile jog (i.e., the step of dislocation unable to glide) is produced after the intersection, the motion of the dislocation is severely impeded, leading to significant strengthening of the materials.

5.15 EXTENDED DISLOCATIONS

On the basis of energy considerations, it is possible that some dislocations may dissociate (split) into shorter segments. Consider an edge dislocation consisting of two half-planes of atoms in a {100} orientation with Burgers vector b_1 being equal to $2a$. Assume that they would separate to form two dislocations with the Burgers vectors, b_2 and b_3, of only one-half of the original dislocation. The dissociation reaction is represented as

$$\vec{b_1} \rightleftharpoons \vec{b_2} + \vec{b_3} \tag{5.74}$$

The free-energy change can be used to determine the occurrence of the reaction. Moreover, to a good approximation, the free energy of a dislocation can be equated to its strain energy per unit length. Thus

$$\begin{aligned} \Delta G &= G_2 + G_3 - G_1 \\ \Delta G &= \mu b_2^2 + \mu b_3^2 - \mu b_1^2 = \mu(b_2^2 + b_3^2 - b_1^2) \\ &= \mu(a^2 + a^2 - 4a^2) = -2\mu a^2 < 0 \end{aligned} \tag{5.75}$$

Note that $b_1 = 2a$; b_2 and $b_3 = a$. Because of the negative value of ΔG, it is expected that the dislocation composed of two extra half-planes would dissociate into two dislocations with a single extra half-plane. This infers that a dislocation with a Burgers vector of na would decompose into n dislocations with each having Burgers vectors a. Extending the argument further, one might expect that the dislocation having one extra plane of atoms and a Burgers vector a would further decompose somehow into two dislocations with each possessing a Burgers vector of less than a, say 1/2 a. If n is not an integer, the passage of the dislocation will produce a plane fault in the crystal structure right on the glide plane. As a consequence, this reaction may produce a combination of the two partial dislocations and a stacking fault called an extended dislocation, represented in Fig. 5.20.

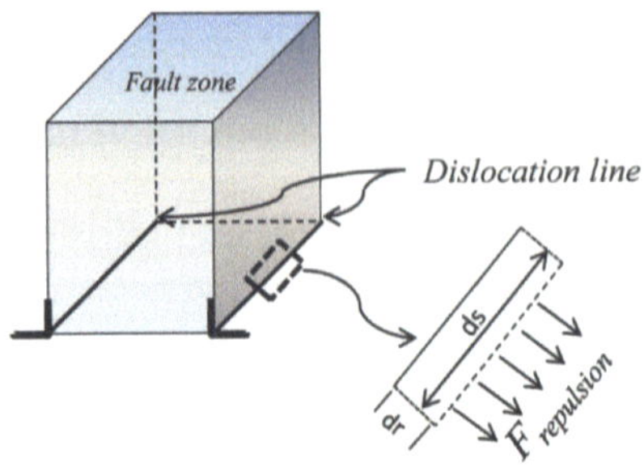

FIGURE 5.20 Schematic presentation of an extended dislocation with two partial dislocations and stacking fault, in which a segment of a two-dimensional fault on the slip plane extended outward for a distance *dr* by a repulsion force per unit length is also presented.

5.15.1 Partial Dislocations

Partial dislocations are significant in twinning reactions, phase transformation, and the formation of dislocation barriers by intersecting dislocations. For an adequate study of deformation and strengthening of metals, we must distinguish between total (or perfect) dislocations and partial (or imperfect) dislocations. The passage of a total dislocation through a crystal leaves the perfection of the crystal unimpaired. Passage of a partial dislocation shifts an atom only a fractional amount of the lattice constant and therefore leaves behind a planar region of imperfect crystal. Moreover, the Burgers vector of a partial dislocation is expressed as that of a perfect dislocation, but it is not a unit lattice vector. Thus, as the Burgers circuit must start and finish in the surface of the stacking fault, the final position of the circuit in the perfect lattice is not a lattice point.

In general, single crystals deform through the slip in a close-packed direction on close-packed planes, in which the slip direction is associated with Burgers vector. The Burgers vector is associated with a lattice vector whose magnitude is the smallest, i.e., the distance between two nearest neighboring lattice points. Essentially, as the magnitudes of the translation vectors of the crystal structure are defined to have unit length, the lattice vectors can be determined only if the coordinates of the lattice points connecting the lattice vectors are known. For convenience, Burgers vectors are further represented by the Miller indices of its direction with a scalar preceding the direction. Usually, in the cubic system, the lattice constant is present, indicating that the magnitude of the components of Burgers vector is expressed in terms of it. However, in order to directly extend to other crystal systems, the lattice constant is omitted. Some examples are illustrated in Fig. 5.21A. Dislocations with such Burgers vectors are considered to be perfect dislocations. Concerning partial dislocations, the end points of their Burgers vectors are not the lattice points. Thus we have to get around this problem by finding out the fraction of the magnitude of the Burgers vector in proportion to the length of the lattice vector along the direction of the Burgers vector.

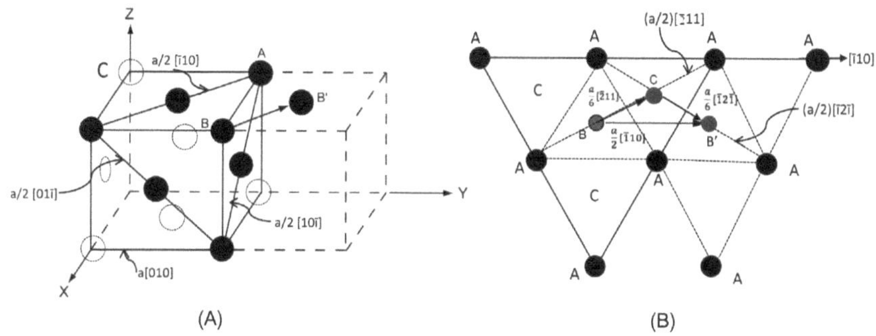

FIGURE 5.21 (A) Representation of the vector notations of different crystallographic directions. (B) Schematic illustration of the orientation relationship of the Burgers vectors of a total dislocation and its dissociated partial dislocations.

In the following, a typical example concerning the slip process occurring in a slip system of $\{111\}\langle 110\rangle$ in FCC metals will be illustrated. In considering the slip occurring on (111) planes and in (a/2)[$\bar{1}$10] directions, where the atom at position B moves to B′, as shown in Fig. 5.21A, two slip processes may occur in view of the energy consideration, namely, one is normal movement for the total dislocation with Burgers vector of $\vec{b}_t = (a/2)[\bar{1}10]$ moving a unit from B to B′ (Fig. 5.21A) and another is zigzag movement including two partial dislocations with Burgers vectors of $(a/6)[\bar{2}11]$ and $(a/6)[\bar{1}2\bar{1}]$, commonly called Shockley partial dislocations, as shown in Fig. 5.21B, in which the A layer is redrawn, based on Fig. 5.21A, and the corresponding B and C sites are indicated. The Burgers vectors of the partial dislocations can be ready to be determined. The magnitude of the Burgers vector from the B site to the nearest C site in the plane is 1/3 of the length of the lattice vector $(a/2)[\bar{2}11]$ along the line A-B-C-A in the plane. Thus the Burgers vector of the first partial dislocation is $(a/6)[\bar{2}11]$. Likewise, the magnitude of the Burgers vector from the C site to the nearest B′ site in the plane is 1/3 of the length of the lattice vector $(a/2)[\bar{1}2\bar{1}]$ along the line A-C-B′-A in the plane. Thus the Burgers vector of the second partial dislocation is $(a/6)[\bar{1}2\bar{1}]$.

By approximating the strain energy as μb^2 and substituting the appropriate Burgers vectors, the energy of the former total dislocation of $\mu a^2/2$ is higher than that of the latter two partial dislocations of $\mu a^2/3$. Therefore, the total Burgers vector tends to dissociate into two partials,

$$\frac{a}{2}\left[\bar{1}10\right] \rightarrow \frac{a}{6}\left[\bar{1}2\bar{1}\right] + \frac{a}{6}\left[\bar{2}11\right] \tag{5.76}$$

5.15.2 Faults in the Stacking in FCC Crystals

As observed in Fig. 5.20, the atoms lying between the two partial dislocations have stacking positions that differ from those of the atoms on the far side of either of the partial dislocations. The discontinuities occur in the stacking order of the close-packed planes, called a stacking fault. Since the stacking faults per se are the barriers to dislocation movement, and they have definitely been observed in FCC structures, the possible development of them will be described. Assume a Shockley partial dislocation gliding along the plane between A and B of Fig. 5.21B, and also assume that the B-layer plane and all planes above it are displaced by the vector $\vec{b}_s = (a/6)[\bar{2}11]$, where B→C, C→A, and A→B, relative to positions fixed on the original A-layer plane, giving ABC**A**C**A**BCABC. The resultant fault arises in two twin planes with bold **A** and **C**, and also it can be considered that one atomic layer is missed between them. It is observed that the normal stacking sequence is maintained in the crystal on either side of the fault right up to the fault plane, so this fault is called an intrinsic stacking fault. By contrast, if the C-layer plane below A and all successively lower planes are displaced by $\vec{b}'_s = (a/6)[2\bar{1}\bar{1}]$, where C→B, B→A, and A→C, resulting in BCA**B**A**C**ABCAB. This fault contains two twin planes of bold **B** and **C**, with one extra pane inserted between them, which is called an extrinsic or double stacking fault, defined as planes inserted at the center of the fault being incorrectly stacked with respect to the layers on both sides of the fault.

On account of the presence of the stacking fault as the Shockley partial dislocation moves, Eq. (5.76) should be rewritten as

$$\frac{a}{2}[\bar{1}10] \rightarrow \frac{a}{6}[\bar{1}2\bar{1}] + \frac{a}{6}[\bar{2}11] + \text{fault}$$

Supposing that the distance between the partial dislocations is r_e, the feasibility for the formation of the extended dislocation depends on the strain energy of dislocations and the stacking fault energy. Let the energy per unit area of the fault be given as E_f. The free energy difference of the reaction is expressed as

$$\begin{aligned}\Delta G &= \mu b_s^2 + \mu b_s'^2 - \mu b_t^2 + E_f r_e \\ &= \mu a^2/3 - \mu a^2/2 + E_f r_e \\ &= -\frac{1}{6}\mu a^2 + E_f r_e\end{aligned} \tag{5.77}$$

Note that ΔG is per unit length here. Accordingly, the reaction would occur when $\Delta G < 0$, namely, $E_f r_e < \frac{1}{6}\mu a^2$, which implies that the formation of the extended dislocation depends on the stacking fault energy and equilibrium distance between partial dislocations.

5.15.3 Equilibrium Separation Between Partial Dislocations

As aforementioned in Section 5.13, dislocations with the same sign at the same slip plane will repulse each other. However, as the dislocations move, the fault area will also be expanded, leading to the increase of the energy and tending to pull the dislocations back. Thus there is equilibrium separation between partial dislocations. There are two ways to solve this issue: energy and force balance viewpoints. For the former, take a small segment of *ds* of the dislocation line, exerted by the repulsive force $F_{\text{repulsion}}$, shown in Fig. 5.20, and it can be expressed in an approximate general form in the following:

$$F_{\text{repulsion}} = \frac{F}{l} \cong \frac{\mu\left(\vec{b_s} \cdot \vec{b'_s}\right)}{2\pi r} \tag{5.78}$$

where l represents the length of the dislocation line, $\vec{b}_s$ and $\vec{b}'_s$ indicate the Burgers vector of each partial dislocation, and the values of the dot product of the Burgers vector depend on the structure. This repulsive force is counterbalanced by an attractive force that results from the faulted region generated between the partial dislocations as they move apart.

Let the faulted region have a surface tension γ in units of dynes/cm. The surface tension is defined as the work necessary to generate a unit area of the fault plane, expressed as

$$\gamma = \frac{dw}{dA}. \tag{5.79}$$

Consider a length, *ds*, along the boundary of the faulted area as shown in Fig. 5.20. The work needed to extend this length outward by a distance *dr* is

$$dw = \frac{F}{l} ds \cdot dr \tag{5.80}$$

At equilibrium, taking the equilibrium distance, r_e, into consideration, Eq. (5.78) becomes

$$\frac{F}{l} \cong \frac{\mu\left(\vec{b_s} \cdot \vec{b'_s}\right)}{2\pi r_e} \tag{5.81}$$

This is the force per unit length exerted on the boundary to move it outward. Thus we obtain

$$\begin{aligned} \frac{F}{l} &= \frac{dw}{ds \cdot dr} = \gamma = \frac{\mu\left(\vec{b_s} \cdot \vec{b'_s}\right)}{2\pi r_e} \\ \gamma &= E_f = \frac{\mu\left(\vec{b_s} \cdot \vec{b'_s}\right)}{2\pi r_e} \\ r_e &= \frac{\mu\left(\vec{b_s} \cdot \vec{b'_s}\right)}{2\pi E_f} \end{aligned} \tag{5.82}$$

Here, it is assumed that the surface tension can be approximated as energy per unit area (see detail in Section 6.3 of chapter 6), and that $\gamma = E_f$.

With respect to the force balance viewpoint, since the two Burgers vectors of the partials are 60° to each other, these partials can be further dissolved as edge and screw components with Burgers vectors of $b_e = b/2$ and $b_s = (b/2)\tan 30°$, respectively (b: Burgers vector of total dislocation). In equilibrium, the net force on these partials must be zero.

$$F = F_e + F_s + F_\gamma = 0 \tag{5.83}$$

where F_e and F_s are of edge (repulsion due to the same sign) and screw (attraction due to one being right-handed and the other left-handed) components, respectively, and F_γ is a surface stress arising from stacking fault (attraction), assumed to be equal to γ. Applying the relationships of the interaction forces between edge and screw dislocations derived in Section 5.13 to Eq. (5.83), obtains:

$$F = 0 = \frac{\mu b_e{}^2}{2\pi(1-\upsilon)r_e} - \frac{\mu b_s{}^2}{2\pi r_e} - \gamma \tag{5.84}$$

Rearranging Eq. (5.84), gives

$$r_e = \frac{1}{\gamma}\left(\frac{\mu b_e{}^2}{2\pi(1-\upsilon)} - \frac{\mu b_s{}^2}{2\pi}\right) \tag{5.85}$$

Substituting $b_e = b/2$ and $b_s = (b/2)\tan 30°$ into Eq. (5.85), yields

$$r_e = \frac{1}{\gamma}\frac{\mu b^2}{8\pi(1-\upsilon)}\left(1-(1-\upsilon)\tan^2 30°\right) \tag{5.86}$$

Letting υ be 1/3, we obtain

$$r_e = \frac{7}{72}\frac{1}{\gamma}\frac{\mu b^2}{\frac{2}{3}\pi} \tag{5.87}$$

Consider a metal with high stacking fault energy such as Al:

$$\gamma \simeq 200 \text{ ergs/cm}^2$$
$$b \simeq 2.8 \times 10^{-8} \text{ cm}$$
$$\mu \simeq 4 \times 10^{11} \text{ dynes/cm}^2$$

$$r_e \simeq \frac{7}{72}\frac{1}{200}\frac{(4\times 10^{11})(8\times 10^{-16})}{\frac{2}{3}\pi}\text{cm} \simeq 7.4 \times 10^{-8}\text{cm}$$

This result indicates that the equilibrium separation distance is extremely small for materials with high stacking fault energy. Thus it is suggested that extended dislocation would not be developed for materials with high stacking fault energy.

In fact, the extended dislocation has a very significant influence on the plastic behavior of FCC metals. For the deformation process, dislocation usually can continue to move via the intersecting slip plane, so-called cross slip, which would be easier for screw dislocations but more difficult for edge dislocations to move on. However, if the screw dislocations dissociate into extended dislocations, the formed Shockley partial dislocations are not pure screw, which would make the cross slip difficult. In addition, cross slip requires that the extended dislocations contract into perfect dislocations before moving onto a new plane via cross slip. Since the wider the separation between partial dislocations, the higher the energy required to contract them, cross slip would become more difficult for the presence of extended dislocation, and in turn for the materials with low stacking fault energy. In addition, the motion of partial dislocations is restricted to the plane of their fault. If the dislocations are able to move by pure glide, they are called glissile dislocations, but cannot move by glide whereas they can climb, called sessile dislocations. The Shockley partial dislocation would be classified as a glissile dislocation because its fault plane is also its glide plane, and the Burgers vector of the dislocation lies in the plane of the fault. Nevertheless, there is another partial dislocation in FCC metals that is sessile dislocation, called Frank partial dislocation. Regarding the stacking faults, if a stacking fault terminates inside a crystal, partial dislocation will be developed at its boundaries, which in general, may be either of the Shockley type, with the Burgers vector of the dislocation lying in the plane of the fault, or of the Frank type, with the Burgers vector normal to the stacking fault. As aforementioned, the motion of the Shockley partial dislocation can produce the intrinsic stacking fault, which may also be formed by removing a layer of atoms from the normal sequence, which is physically possible to occur by the condensation of vacancies on an octahedral plane. However, it is noticed that this fault in this manner is similar to that arising from the slip of a Shockley partial dislocation, but the partial dislocations at the boundaries of the fault are not the same, these are normal to the $\{111\}$ slip plane and thus of the Frank type whose Burgers vector is $1/3\langle 111\rangle$, shown in Fig. 5.22. The Frank partial dislocations basically form a loop running around the outer boundary of the vacancy disk region.

On the contrary, the extrinsic or double stacking fault formed by the slip of a Shockley partial dislocation can be generated by the precipitation of interstitial atoms on an octahedral plane. The created partial dislocation is also Frank type with the Burgers vector $1/3\langle 111\rangle$, shown in Fig. 5.23.

As observed in Figs. 5.22 and 5.23, it is obvious that the slip planes of these Frank partial dislocations are not the fault planes, so Frank partial dislocations are sessile dislocations.

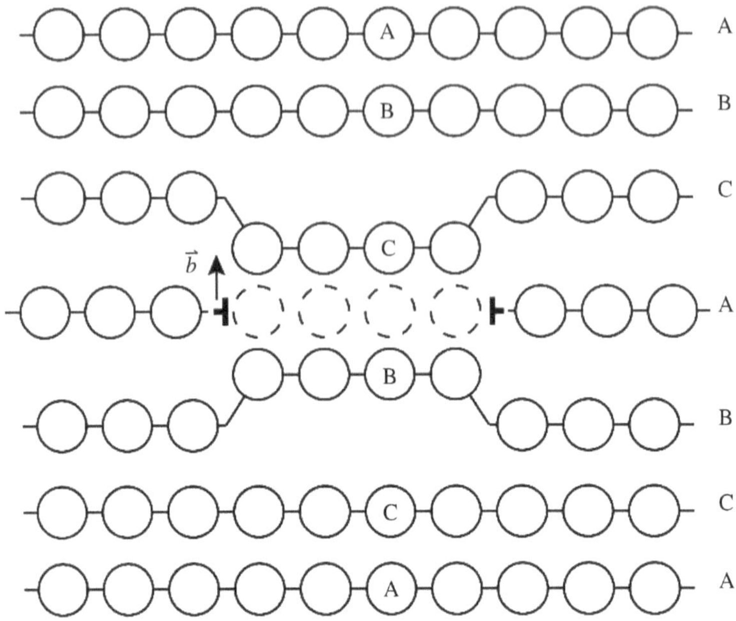

FIGURE 5.22 Illustration of the intrinsic stacking fault formed in FCC crystal by removing a layer of atoms from the normal sequence, which is physically possible to occur by the condensation of vacancies on an octahedral plane.

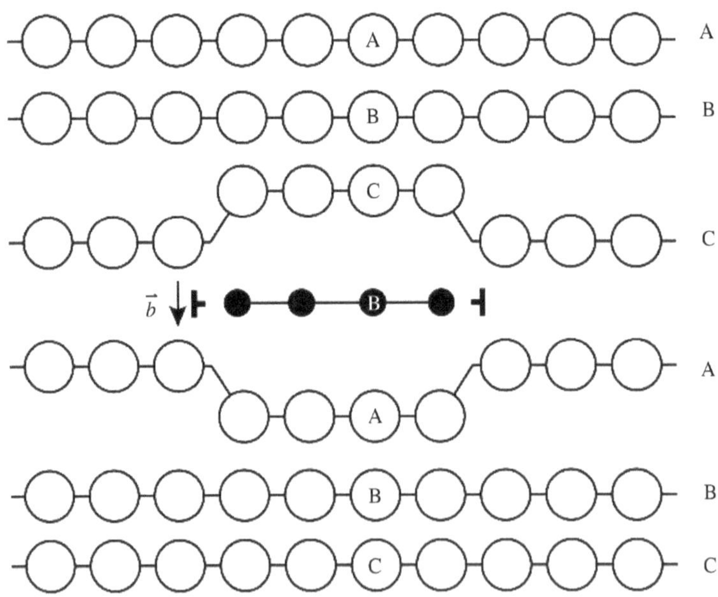

FIGURE 5.23 Illustration of an extrinsic or double stacking fault produced by inserting an extra plane in a FCC crystal, which physically could be formed by the precipitation of interstitial atoms on an octahedral plane.

Essentially, the development and the related properties of the extended dislocations in other crystal systems are analogous to those of FCC crystals; whereas the partials of HCP and BCC crystals may differ from the FCC structure. The extended dislocation occurs in hexagonal metals on the basal plane via the following dissociation reaction:

$$\frac{1}{3}\left[\bar{1}2\bar{1}0\right] = \frac{1}{3}\left[01\bar{1}0\right] + \frac{1}{3}\left[\bar{1}100\right] + \text{fault} \tag{5.88}$$

Chapter 6

Two-Dimensional (Interfaces) and Three-Dimensional (Second Phases) Imperfections in Solids

Interfaces are two-dimensional imperfections that serve as boundaries between one orderly region of a crystal and another. The interface is commonly a transition region in which the order of the crystal is interrupted. The ordered states may refer to the electronic state, the magnetic state, the chemical composition, or the configurational arrangements of atoms. Typical boundaries in the materials are categorized as follows:

1. Free surface: Interface between solids and gases.
2. Domain boundaries: Interface between regions where there is a change in the electronic structure but no change in the periodicity of atom arrangement.
3. Grain boundaries: Interface between two crystals or grains of the same phase where there is an orientation different in the atom arrangement across the interface.
4. Interfacial (phase) boundaries: Interfaces between different phases where there is generally a change of chemical composition and atom arrangement across the interface.

6.1 GRAIN BOUNDARIES

In practice, virtually all the materials used daily are polycrystalline. One of the goals of material science and engineering is to gain control over microstructure in the synthesis and processing of materials and thus to produce material with outstanding performance. Manipulation of the grain boundaries is a key issue in determining the material properties. Hence, understanding of the grain-boundary structure is necessary. For simplicity, a two-dimensional grain boundary will be taken into account, which can be completely defined by specifying the angles with respect to the orientation of one lattice with respect to the other and the orientation of the boundary with respect to a lattice. Note that a boundary specified with two angles is called a two-degree-of-freedom boundary.

Elements of Structures and Defects of Crystalline Materials.
DOI: http://dx.doi.org/10.1016/B978-0-12-814268-4.00006-0

For the grain boundary in a three-dimensional lattice, consider the crystal shown in Fig. 6.1A. Assume that this crystal is cut on the $x-z$ plane and is rotated about the three axes to form an orientation mismatch between two crystals, wherein the tilt boundaries are formed for the x and z axes but a twist boundary is formed for the y axis. Further, take the tilt boundary formed via the rotation of x axis as the reference, and rotate the boundaries about the x and z boundary axes. The resultant boundary positions are represented in Fig. 6.1B.

It is implied that in order to specify the orientation of two crystals, one must specify three angles by means of the rotation about the x, y, and z axes (Fig. 6.1A). Additionally, concerning the boundary between two crystals of fixed orientation, shown in Fig. 6.1B, since the boundary lies in the $x-z$ plane, the boundary plane could be changed by rotating it about either x or z but not y. Hence, to specify a boundary orientation between two crystals we must specify two angles. Therefore, for three-dimensional lattices, the general grain boundary has five degree of freedom. Three degrees specify the orientation of one grain relative to the other and two degrees specify the orientation of the boundary relative to one of the grains. Furthermore, if the rotation angle, ω, is contained in the boundary, the grain boundary is a tilt boundary (the simplest edge dislocation grain boundary), while for the y axis,

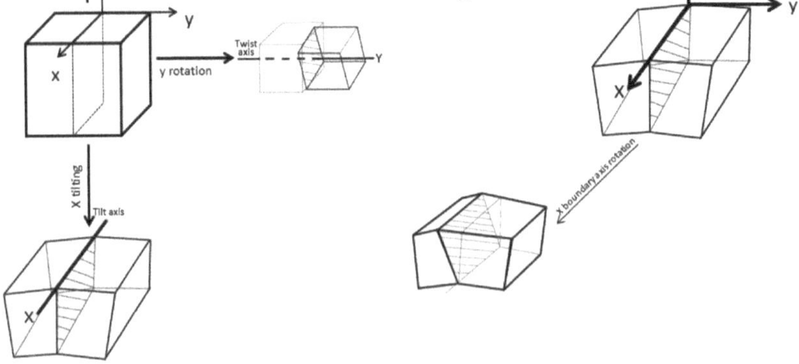

FIGURE 6.1 Schematic presentation of the formation of the five degrees of a grain boundary in a three-dimensional lattice: (A) Take a perfect crystal cut into two, and tilt along the x and z axes but rotate the y axis. (B) Take the crystal with a boundary tilted along the x axis, and rotate this boundary along the x and z axes.

ω is perpendicular to the boundary, it is a pure twist boundary (the simplest screw dislocation grain boundary).

Grain boundaries are often classified by a scheme referring to a one-degree-of-freedom boundary based on geometry concerning orientational mismatch or the degree of difference in orientation:

1. Subboundaries within a given grain.
2. Small-angle boundaries $\theta = 0 \rightarrow 0-10$ degree.
3. Medium-angle boundaries $\theta = 3-10 \rightarrow 0$ degree.
4. Large-angle boundaries $\theta = 15$ degree$\rightarrow$.

Grain boundaries can also be distinguished by the type of difference in orientation: (1) tilt boundaries, where the misorientation can be described as a relative tilting between two lattices; and (2) twist boundaries, where the misorientation can be described as a relative rotation between two lattices.

6.1.1 Tilt Boundary

The symmetrical tilt boundary in a simple cubic crystal can be pictured by considering the lattice points of two grains, initially in the same orientation, symmetrically titled about the x axis; as a result, each (010) plane from one grain to another grain ends as an incomplete plane in the boundary at alternating intervals, yielding a tilt boundary that consists of a succession of elementary edge dislocations with Burgers vector b, shown in Fig. 6.2.

The relationship of the spacing between dislocations and tilt angle can be modeled by assuming that the edge dislocations pile up along the (010) plane, illustrated in Fig. 6.2 (right). It is further assumed that the lattice constant is approximated as the magnitude of the Burgers vector, b, and the selected tilt angle δ can generate a uniform array of the edge dislocations,

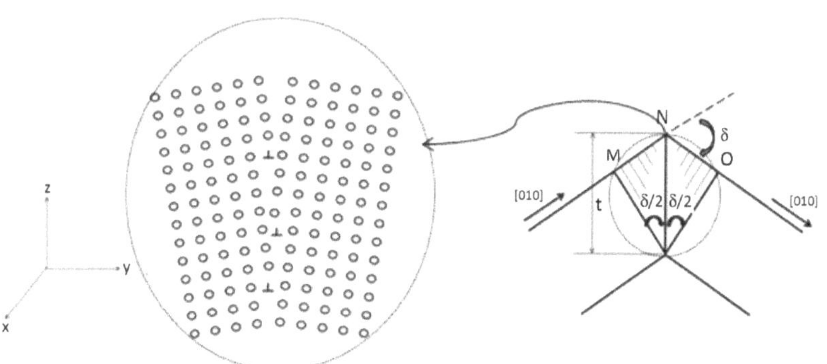

FIGURE 6.2 The feature of a symmetrical tilt boundary in a simple cubic crystal, wherein the two lattices can be maintained continuous by inserting an edge dislocation periodically, in which the relationship of the spacing between dislocations and tilt angle concerning a (010) tilt boundary is modeled on the right.

and thus the separation d between the edge dislocations is fixed. All (010) planes intersecting the surface within the region MNO are considered to end as incomplete planes at the boundary. The number of the incomplete planes can be obtained:

$$n = \frac{2t}{b}\sin\frac{\delta}{2} \tag{6.1}$$

where n is the incomplete planes and b is Burgers vector. The average separated distance, d, between dislocations in the boundary is

$$d = \frac{t}{n} = \frac{b}{2\sin\frac{\delta}{2}} \tag{6.2}$$

When δ is taken to be small,

$$\sin\frac{\delta}{2} \cong \frac{\delta}{2} \tag{6.3}$$

$$d \cong \frac{b}{\delta} \tag{6.4}$$

Eq. (6.4) can further be used to correlate the tilt angle with the grain boundary energy, which has a significant influence on the equilibrium grain shape and the grain boundary mobility. Since the tilt boundary is constructed by the pile-up dislocations, its energy would vary with the change of δ because the number of dislocations varies with δ/b according to Eq. (6.4). Moreover, as per Eq. (5.32), the energy of each dislocation in the tilt boundary can be expressed as:

$$E_e = \frac{\mu b^2}{4\pi(1-\upsilon)}\ln\frac{d}{2r_0} \tag{6.5}$$

Due to the fact that $d \sim b/\delta$, the boundary energy also depends on the tilt angle. Let $r_0 = 2b$ and E_c be the energy per unit length of the dislocation core, the energy of the tilt boundary is then given by the product of the strain energy per unit length of the dislocation multiply the density of the dislocations (1/d):

$$E_b \simeq \frac{\delta}{b}\left[E_c + \frac{\mu b^2}{4\pi(1-\upsilon)}\ln\frac{1}{4\delta}\right] \tag{6.6}$$

On account of the assumption of the very small angle, Eq. (6.6) actually only is valid with the angles below about 2–3 degree even if it can be fit for larger angles.

Nevertheless, the dislocation model concerning the low-angle grain boundaries essentially can be extended to the high-angle grain boundaries, wherein the energy is expected to increase slightly due to the increase of the density of dislocation cores. Moreover, the uniform array of the edge dislocations exists only for a few tilt angles, which occurs when $\cot(\delta/2) = 2n$, where n is an integer for a simple cubic crystal. For instance, $\delta = 8.2$ degree, in which there is one dislocation on every seventh cube plane; particularly, as n is one, $\delta = 53$ degree, whose structure forms a twin plane. These angles would be the cusps appearing in the plot of the grain boundary energy versus misorientation angle for symmetrical tilt boundaries. In addition, a typical tilt boundary usually is asymmetric, namely, the plane of the boundary is not consistent with the symmetric orientation, or the total angle of misorientation is not divided equally between grains 1 and 2.

6.1.2 Twist Boundary

A pure twist boundary can be considered as arising from the rotation of one grain relative to its neighbor about an axis perpendicular to the plane of the boundary, shown in Fig. 6.1A (y rotation). Thus the simple twist boundary always requires at least two sets of dislocations for its representation. The corresponding displacements are equivalent to two separate shears at right angles, wherein each shear can be produced by a set of parallel screw dislocations. Consequently, two such sets of parallel screw dislocations orthogonal to each other with the same mean separation for the same magnitude of Burgers vector form a crossed grid of screw dislocations (Read, 1953).

6.1.3 Coincidence Site Lattice

Owing to the fact that the atoms in the boundary do not fit perfectly for the lattice of either adjacent grain, the grain boundary would be a region of relatively high energy. Nevertheless, there are certain relative orientations of two grains, where a certain fraction, $1/n$, of the atomic sites in both grains have corresponding locations. These special sites indicate a coincidence lattice. As shown in Fig. 6.3, the grain 1 adjacent to grain 2 is considered to be rotated 37 degree relative to grain 2. An array of coincidence sites extends uniformly from grain 1 into grain 2, unaffected by the existence of the grain boundary.

As observed in Fig. 6.3, one atom out of five is a coincidence site, and the fraction of atoms in a coincidence at a boundary is generally known as the density of coincidence sites. Nevertheless, the reciprocal of the density is more widely used as a parameter to describe a coincidence site boundary, designated by the Greek symbol Σ, and $\Sigma = 5$ in this case. Actually, coincidence site lattice is one of the more advanced concepts relating to the structure of high-angle grain boundaries. Another is referring to as the structural unit model for grain boundary structure wherein a small number of specific polyhedral arrangements of atoms can be identified from which the atomic structure of the grain boundary region can be constructed for a wide range of boundary misorientations.

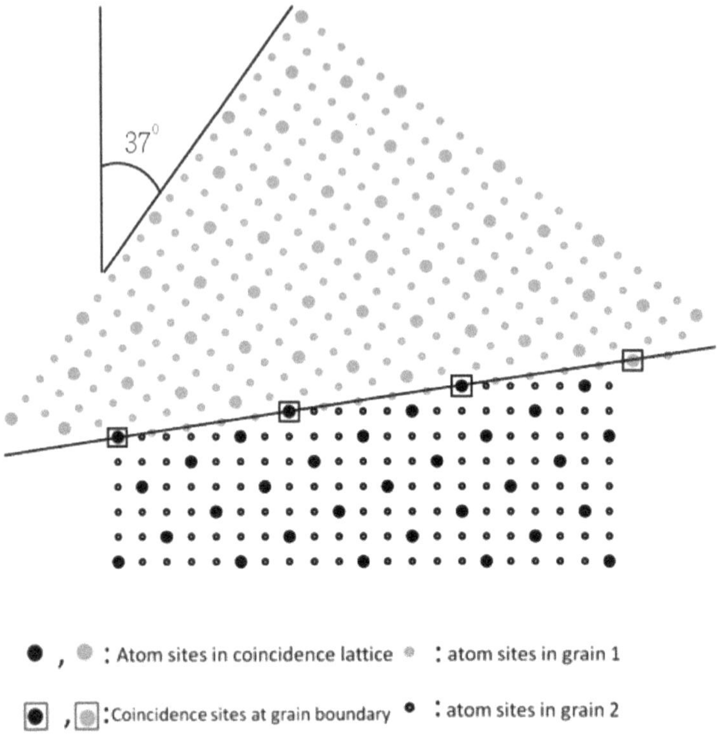

FIGURE 6.3 Schematic illustration of a coincidence lattice in a two-dimensional square lattice, where a 1/5 coincidence lattice presented as two adjacent lattices misoriented by 37 degree.

6.2 INTERPHASE BOUNDARIES IN SOLIDS

In this section, we are dealing with phase boundaries in solids where the two adjacent crystals can have different structures and/or compositions. Because of the different crystal structures, each phase can have a different lattice constant. Thus an interphase boundary can be divided into three categories, coherent, partially (or semicoherent), and incoherent, depending on the degree of fit of the lattices across the boundary. A description of atomic matching must include both the planes (*hkl*) across which matching presents, and the directions [*hkl*] in the matching planes along which the atoms are in registry.

6.2.1 Coherent Boundary

As two phases abutting an interphase boundary have closely related crystal structures, it is expected that there would be a good degree of matching of lattices. If there is one to one matching of the lattice planes across the boundary, the interphase boundary is said to be coherent, shown in Fig. 6.4.

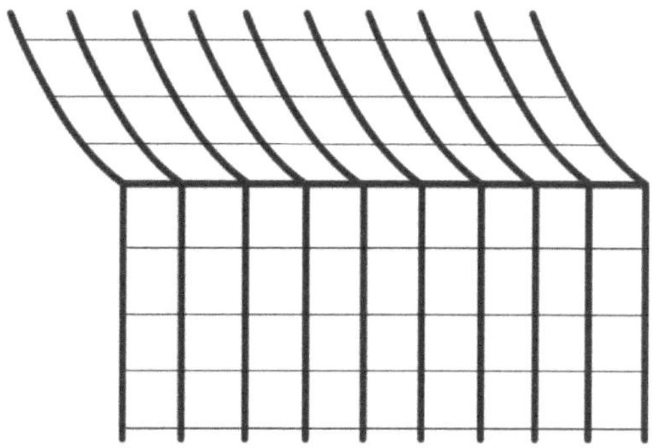

FIGURE 6.4 A coherent phase boundary.

In order to maintain one to one matching, the lattice planes have to be bent, which in turn would produce lattice strains around the boundary, known as coherency strains.

6.2.2 Incoherent Boundary

An interphase boundary in the two adjacent phases having no regularity of lattice plane matching across the boundary is called an incoherent (incommensurate) boundary whose bonding across the interface is sufficiently poor that there is no local relaxation of the misfit into dislocation at the interface. Thus the atomic structure is very little known, but it has many features in common with high-angle grain boundaries.

6.2.3 Partially Coherent or Semicoherent Boundary

An elastic strain is associated with a coherent boundary increasing the total energy of the system. Therefore, for sufficiently large lattice misfit, it becomes energetically more favorable to replace the coherent boundary with a semicoherent boundary, wherein the periodic interfacial dislocations are introduced to accommodate the difference in lattice constants of the phases and thereby reduce the elastic strain energy of the interface. Consider two cubic lattices abutting on their respective planes, as shown in Fig. 6.5.

The unstressed interplanar spacings of matching planes in the phases 1 and 2 are a_1 and a_2, respectively, that are taken so that $a_1 > a_2$. The disregistry or misfit, δ, between the two lattices is defined as

$$\delta = \frac{a_1 - a_2}{a_1} \tag{6.7}$$

It can further be proved that the lattice misfit can be accommodated by a set of edge dislocations with a spacing D given by

$$D = a_2/\delta \tag{6.8}$$

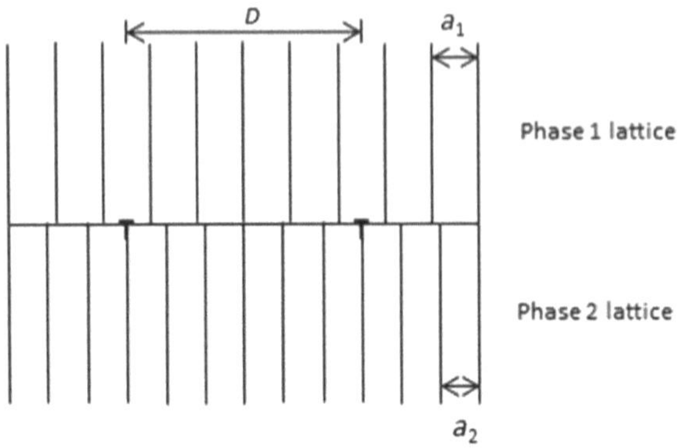

FIGURE 6.5 A partially coherent phase boundary.

The matching in the interface is now nearly perfect except around the dislocation core where the structure is highly distorted and the lattice planes are continuous.

Concerning the energy of interphase boundaries, a coherent boundary is the lowest, a partially coherent (or semicoherent) boundary intermediate, and an incoherent boundary the highest. Nevertheless, the stress field corresponding to the high strain energy of the coherent boundary could be more effective in stopping dislocations.

6.3 SURFACE TENSION, SURFACE STRESS, AND SURFACE FREE ENERGY OF INTERFACES

Since all materials are finite and would contact with their environment at either a free surface or an interface with another phase, they all possess surface imperfections. Surface usually refers to the external boundary between solid and a gaseous phase or a vacuum but interfaces are considered to be the internal boundaries between condensed phases. Surface properties are usually quite different from those in the bulk of the phase. Material properties are significantly influenced by the interactions of either free surface or interfaces with other imperfections, either point or line defects.

While surface tension is more widely used in describing the surface imperfection than surface stress and surface free energy, the latter two fundamentally should be more correctly adopted in the related subjects. Underlying a thermodynamic point of view, the surface free energy γ indicates the reversible work per unit area involved in forming a surface, where new atoms are exposed. The surface stress σ is the reversible work per unit area required to stretch a surface.

The relationships of surface stress and surface free energy in liquids or solids can be deduced as follows. For a liquid, the cohesive forces among liquid molecules are responsible for the phenomenon of surface stress. In the bulk of the liquid, each molecule is shared with all neighboring atoms. However, those on the surface have no neighboring atoms above, and exhibit stronger attractive forces upon their nearest neighbors on the surface. This enhancement of the intermolecular attractive forces at the surface is called surface tension or stress. Moreover, the surface is pulled inwards creating some internal pressure and forces liquid surfaces to contract to the minimal area.

Another way to view surface tension is in terms of energy. A molecule in contact with a neighbor is in a lower state of energy than if it were alone (not in contact with a neighbor). The interior molecules have as many neighbors as they can possibly have, but the boundary molecules are missing neighbors (compared to interior molecules) and therefore have a higher energy. For the liquid to minimize its energy state, the number of higher energy boundary molecules must be minimized. The minimized quantity of boundary molecules results in a minimal surface area. As a result of surface area minimization, a surface will assume the smoothest shape it can. Since any curvature in the surface shape results in greater area, a higher energy will also result. Consequently, the surface will push back against any curvature.

Considering a liquid with surface area A, the free energy is

$$G = G_0 + A\gamma \tag{6.9}$$

where G_0 is the free energy in the bulk of the liquid and γ the excess free energy. Assuming that the surface is extended by an area dA, the work done by the surface stress σdA can be considered to increase the free energy of the system dG. Thus

$$dG = \gamma dA + Ad\gamma = \sigma dA \tag{6.10}$$

$$\sigma = \gamma + A\frac{d\gamma}{dA} \tag{6.11}$$

When a fluid surface is stretched, new atoms or molecules arrive at the surface adopting the same structural configuration such that the number of atoms per unit area remains constant. Thus the surface energy of liquid is independent of the area of the interface, i.e.,

$$\frac{d\gamma}{dA} = 0 \tag{6.12}$$

$$\sigma = \gamma \tag{6.13}$$

As a result, it can be considered that for a fluid, the surface free energy is the same as the surface stress.

As far as the solid surfaces are concerned, the origin of the surface stress based on the atomic scale can be perceived in the following argument.

Consider a single-crystal material. Owing to the dangling bonds, electron density of the surface atoms is lower in comparison to the bulk atoms, and therefore the surface atoms adopt an equilibrium spacing different from the bulk. Therefore, bulk atoms would impose the constraint that the surface atoms are forced into an epitaxial relationship with the underlying lattice. In this sense, the surface atoms are strained, and the surface has a stress exerted on it by the underlying lattice.

A planar surface of solids can be created for instance, either by plastic deformation or cleavage. The total work needed to create a planar surface of area A (equivalently, the total excess free energy of the surface) is equal to γA. Now the so-called surface stress associated with the reversible work per unit area would be considered to elastically stretch a preexisting surface. As aforementioned, the surface atoms are strained. Hence, a surface elastic strain tensor, ε_{ij}, is assumed to represent the elastic deformation of a solid surface. Consider a reversible process that causes a small variation in the area through an infinitesimal elastic strain $d\varepsilon_{ij}$. The surface stress tensor σ_{ij} can further be defined as the work associated with the variation in the total excess free energy of the surface (γA), on account of the strain $d\varepsilon_{ij}$:

$$d(\gamma A) = A\sigma_{ij}d\varepsilon_{ij} \tag{6.14}$$

Since $d(\gamma A) = \gamma dA + Ad\gamma$, and $dA = A\delta_{ij}d\varepsilon_{ij}$ (where δ_{ij} is the Kronecker delta and is 1 for $i = j$ but 0 for $i \neq j$), the surface stress σ_{ij} can be expressed as

$$\sigma_{ij} = \gamma\delta_{ij} + \partial\gamma/\partial\varepsilon_{ij} \tag{6.15}$$

Thus the state of stress on a surface is given as

$$\sigma_{xx} = \gamma + \partial\gamma/\partial\varepsilon_{xx}, \tag{6.16}$$

$$\sigma_{yy} = \gamma + \partial\gamma/\partial\varepsilon_{yy}, \tag{6.17}$$

and

$$\sigma_{xy} = \partial\gamma/\partial\varepsilon_{xy}, \tag{6.18}$$

where ε_{xx} and ε_{yy} are normal strain components and ε_{xy} is the shear component. In addition, the surface stress tensor can be referred to a set of principal axes such that the off-diagonal components are equal to zero. If the surface possesses a threefold or higher rotation axis symmetry, the diagonal components are equal, and the surface stress can become a scalar $\sigma = \gamma + \partial\gamma/\partial\varepsilon$. In general, when a solid surface is elastically stretched, the actual number of atoms per unit area is altered, i.e., $\partial\gamma/\partial\varepsilon \neq 0$, and thus, $\sigma \neq \gamma$. However, for most solids, σ is generally of the same order of magnitude as γ and can be positive or negative. Especially, for high-temperature processes, the distinction between them is often unimportant.

Nevertheless, through dimensional analysis, both surface stress and surface free energy with the unit of Jm^{-2} can be considered as representing a force per unit length (Nm^{-1}), as a result, they would be referred to as surface tension. However, for solid surfaces, σ and γ are fundamentally distinct. The former is exerted by a surface during elastic deformation, and the latter is exerted by a surface during plastic deformation. Therefore, it is suggested that it is probably best to avoid using the term of surface tension when discussing solid surfaces.

Another fundamentally important difference between liquid and solid surfaces is relevant to surface energy anisotropy. Without long-range order structures, the surface free energies of liquids are isotropy and thus liquid droplets tend toward a spherical shape. But for solids, their crystal structures possess long-range order forming the so-called the Bravais lattice, wherein planes on which the atoms are close-packed defining a set of possible "low-energy planes." For instance, a cubic crystal may have low-energy planes on the faces of the cube or on the diagonals. The planes are low-energy in the sense that if the crystal is cleaved along these planes, there will be relatively few broken bonds and a relatively small increase in energy over the unbroken crystal. Equivalently, these planes have a low surface free energy and will form the largest facets, in order to minimize the overall thermodynamic free energy of the crystal. Thus the surface free energies of solids have anisotropy along with the faceted surface.

6.4 FREE SURFACE

An excess of energy on a surface is due to differences in the surroundings of atoms at the surface and in the interior, where there are broken or distorted bonds resulting in an increase in energy. Surface free energy can be defined as the increase in free energy per unit area of new surface formed:

$$\gamma = \left(\frac{dG}{dA}\right)_{T,P,n_i} \tag{6.19}$$

Alternatively, surface free energy can also be defined as the energy associated with the chemical bonds that must be broken to generate the surface, which can be specified as the number of broken bonds per unit area of surface formed multiplied by binding energy per bond.

Taking into account (001) and (011) surfaces, the distance between bonds along the (001) surface is a, while it is only $0.707a$ along the (011) surface. Since the number of bonds per unit area is higher along the (011) plane, $\gamma_{011} > \gamma_{001}$. The results indicate that the surface free energy in a crystal has anisotropy and in general, the highest atom density planes have the lowest values of γ.

6.4.1 Calculate the Surface Free Energy of a Pure Solid FCC Metal

There are several assumptions made in calculating the surface free energy of a pure solid FCC metal:

i. The bond energy of each atom merely takes into account the nearest neighbors.
ii. The temperature dependence of energy of each bond (ε_b) can be ignored.
iii. The bond energy of one mole of atoms can be approximated by the heat of sublimation per mole (ΔH_s) of a crystal measured at 0°K and low pressure.
iv. For solids, at low pressure, the enthalpy can be approximated as internal energy.

Regarding the FCC structure, the nearest neighbors (coordination number z) of each atom on a {111} surface are 12, in which there are six atoms in the plane, three above, and three below. As the crystal is cleaved to form two {111} planes, each atom on either of the two surfaces shares three broken bonds. Therefore, in comparing with the atoms in the bulk, every surface atom possesses an excess internal energy of $3\varepsilon_b/2$, represented by E_s in joules per atom:

$$E_s = 3\varepsilon_b/2 \tag{6.20}$$

The bond energy of one mole of atoms is given by

$$E_b = 1/2zN_A\varepsilon_b, \tag{6.21}$$

where N_A is Avogadro's number and the factor of 1/2 arises because we have counted each bond twice in the product zN_A. Since E_b is equal to the heat of sublimation per mole ($\triangle H_s$) at low pressure, we obtain

$$\varepsilon_b = 2\Delta H_s/zN_A \tag{6.22}$$

Combining Eqs. (6.20) and (6.22), gives

$$E_s = 3\Delta H_s/12N_A = \Delta H_s/4N_A \tag{6.23}$$

Thus the surface energy per surface atom on $\{hkl\}$ surface can be written as

$$\gamma = N_{s\{hkl\}}\Delta H_s/12N_A \tag{6.24}$$

where $N_{s\{hkl\}}$ is the number of the broken bonds on $\{hkl\}$ surface.

For {111},

$$\gamma = N_{s\{111\}}\Delta H_s/12N_A = \Delta H_s/4N_A \tag{6.25}$$

For {200},

$$\gamma = N_{s\{200\}}\Delta H_s/12N_A = \Delta H_s/3N_A \tag{6.26}$$

For {220},

$$\gamma = N_{s\{220\}}\Delta H_s/12N_A = 5\Delta H_s/12N_A \tag{6.27}$$

For FCC metals with lattice constant a, the area (A) occupied by each surface atom at the (hkl) plane can be expressed as:

$$A = \frac{a^2\sqrt{h^2 + k^2 + l^2}}{4} \tag{6.28}$$

The surface energy per unit area for each surface can be obtained by dividing Eqs. (6.25), (6.26) and (6.27) with A.

Concerning the broken bonds of {220}, it is in general considered that each surface atom is connected to two nearest neighbors in the {220} plane. Therefore it must be connected to 10 others out of the plane. Since the atoms are symmetrically disposed about the {220} plane, there must be five bonds above the plane of the paper and five below (give a total of 12). However, if the second layer of atoms below the surface would be taken into account, there would be a sixth broken bond at the {220} surface (Howe, 1997). Thus

$$\gamma = \Delta H_s/2N_A. \tag{6.29}$$

Underlying the above results, it is inferred that {111} has the lowest surface free energy.

6.5 INTERFACES OF PHASES

Like free surfaces having an associated surface energy, interfacial boundaries between phases such as liquid–liquid, solid–liquid, and solid–solid are also characterized by an interfacial energy required to create a unit area of a new interface. However, the interfacial energy is always less than the sum of the individual surfaces energy for otherwise the interface would spontaneously separate to lower the total energy. In the following, we will discuss interfacial boundaries of phases in equilibrium.

6.5.1 Wetting

A simple and important case of interfacial equilibrium between phases is that which occurs on a solid surface, involving a liquid phase and a vapor phase. The liquid shape conforms to a minimum of the total interfacial energy under the vapor–liquid–solid equilibrium, shown in Fig. 6.6.

The liquid tends to form a cap with a contact angle θ, defined as shown in Fig. 6.6 and determined by the balance of surface tension of these three interfacial boundary. For $\theta < 90$ degree (Fig. 6.6 (top)), it can be expressed as:

$$\gamma_{SV} = \gamma_{SL} + \gamma_{LV}\cos\theta \tag{6.30}$$

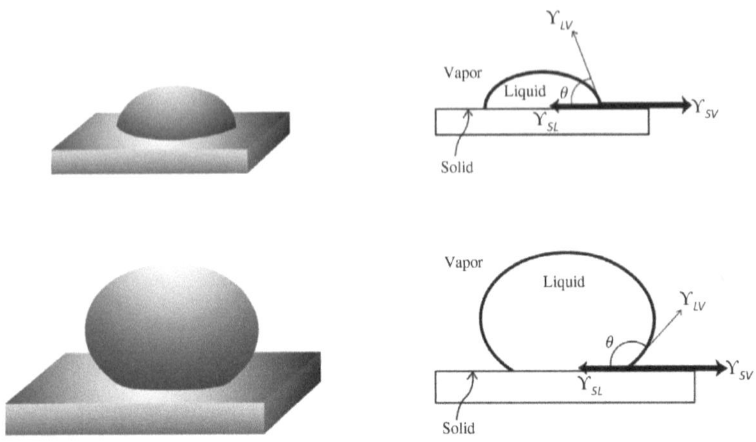

FIGURE 6.6 Schematic illustration of the force balance of the surface tension with a contact angle θ for a liquid drop on a surface.

where γ_{SV}, γ_{LV}, and γ_{SL} are the measured interface energies between the phases present in the system. Regarding $\theta > 90$ degree (Fig. 6.6 (bottom)), $\gamma_{SV} + \gamma_{LV}\cos(\pi - \theta) = \gamma_{SL}$, which actually is equivalent to Eq. (6.30). Thus Eq. (6.30) will be adopted as the general equation, which is further rewritten as

$$\cos\theta = \frac{\gamma_{SV} - \gamma_{SL}}{\gamma_{LV}} \tag{6.31}$$

When $\theta = 180$ degree, the liquid droplet becomes a sphere. For the case where $\theta = 0$ degree, the liquid would completely cover the solid surface, namely, wetting or spreading occurs. At $\theta = 90$ degree, this value is often used to indicate the boundary between nonwetting and wetting behavior. For nonwetting condition,

$$\gamma_{SL} \geq \gamma_{SV} + \gamma_{LV} \tag{6.32}$$

but for wetting condition,

$$\gamma_{SL} \leq \gamma_{SV} - \gamma_{LV} \tag{6.33}$$

Concerning the microstructural development, the grain shape and the second phases present in the grain boundaries would play a significant role in it, which will be discussed in the following.

6.5.2 Equilibrium Shapes of Grains

This section is concerned about the single-phase materials. The microstructures of polycrystalline materials basically are not in a real equilibrium since the free energy of a polycrystal is enhanced due to the high-energy grain boundaries in comparison to the single crystal. Nevertheless, a metastable equilibrium can be reached at the grain-boundary junctions via annealing. The equilibrium conditions at a grain-boundary junction can be obtained either by an energy viewpoint to consider the total grain boundary energy associated with a particular configuration or by considering the forces relating to each boundary exerted on the junction. The detail derivation will be skipped since it is available in many textbooks (e.g., Verhoeven, 1975). If the boundary energy is independent of the orientation, i.e., the torque term (variation of the boundary energy with the change of the orientation) is neglected, the metastable equilibrium at a junction of three grains (Fig. 6.7) can be reached under the following condition:

$$\frac{\gamma_{12}}{\sin\varphi_3} = \frac{\gamma_{13}}{\sin\varphi_2} = \frac{\gamma_{23}}{\sin\varphi_1} \tag{6.34}$$

where φ_i is the dihedral angle between boundaries.

According to Eq. (6.34), if the grain boundary energy is assumed to be the same in a polycrystal, the dihedral angles φ_i would be equal to 120 degree. If the polygons with equal angles are used to represent the grains, the polygon with angle equal to 120 degree is six-sided. Thus the boundary of the grain with six sides can be considered to be planar (Fig. 6.7A). Regarding the sides of a polygon less than six, its angle is smaller than 120 degree. In order to fit in the equilibrium angle 120 degree, the sides become concave inwards (Fig. 6.7A). As for the sides of a polygon more than six, its angle is larger than 120 degree, and the sides become concave outwards (Fig. 6.7A). Accordingly, a grain with the number of boundaries (N) less than six will have curved boundaries and the curvature will be concave inwards, and the N of a grain more than six will be concave outwards. As detailed in Section 6.6, any portion of the grain boundary will be subjected to a normal pressure directed toward its concave. As a result, the former will shrink but the latter will grow during grain growth following the $(N-6)$ rule by Smith (1952). As observed in the micrograph of Fig. 6.7B, grain shapes of most grains are consistent with the aforementioned suggestion. However, the grain shape may not follow the regular polygon principle, due to the anisotropy or impurity effect.

Concerning the microstructural evolution, observation of the change of the grain morphology is very important. Thus grain boundaries shall be manifested by either thermal or chemical etching, wherein grooves would be formed in which grain boundaries intersect surface (Fig. 6.8).

Underlying Fig. 6.8, the force balance of the surface tension (γ_{SV}) and grain boundary surface tension (γ_{SS}) can be written as:

$$\gamma_{SS} = 2\gamma_{SV}\cos\frac{\theta}{2} \tag{6.35}$$

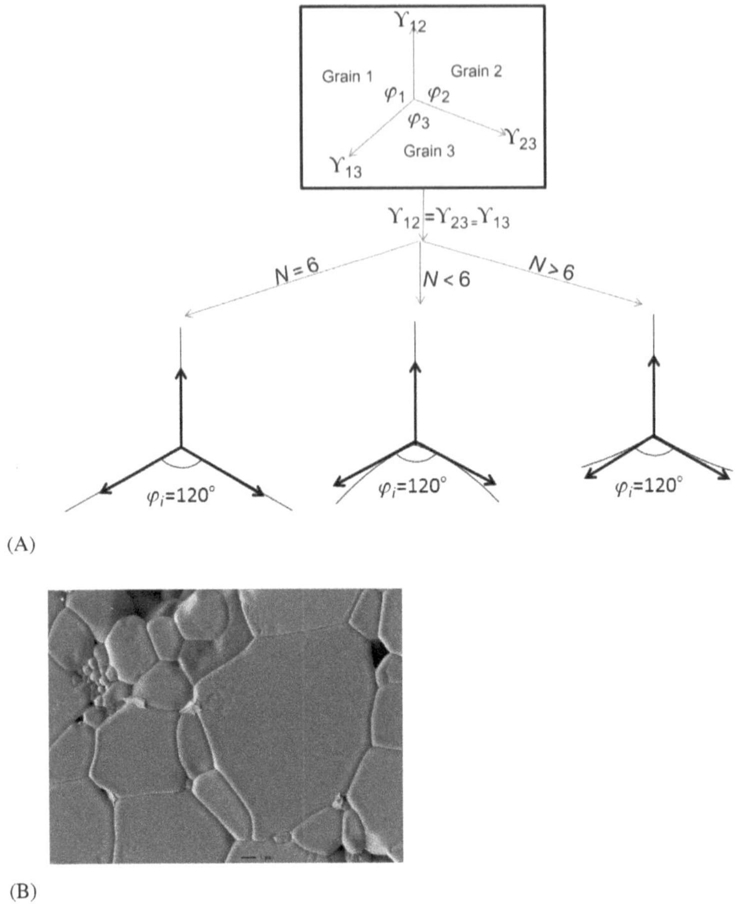

FIGURE 6.7 (A) The grain-boundary surface stress force balance at a junction of three grains. Assume equal surface stress forces, resulting in the dependence of the grain-boundary curvature on the number of the grain boundary (N): equilibrium flat boundary with number equal to 6, concave boundary with number less than 6, and convex boundary with number more than 6. (B) The microstructure of $CaCu_3Ti_4O_{12}$ ceramic, showing that grain shapes of most grains are consistent with the inference of (A).

6.5.3 Morphological Changes of Second Phases

This section essentially is related to the three-dimensional defects including the inclusions and pores. The distribution of the second phases in a polycrystalline material would be located at (1) a corner usually at a junction between four grains, (2) an edge most commonly at a junction between three grains, (3) a face between two grains, i.e., grain boundary, or (4) the interior of a grain, schematically illustrated in Fig. 6.9A. A two-dimensional distribution of the second phases in $CaCu_3Ti_4O_{12}$ has been shown in Fig. 6.9B.

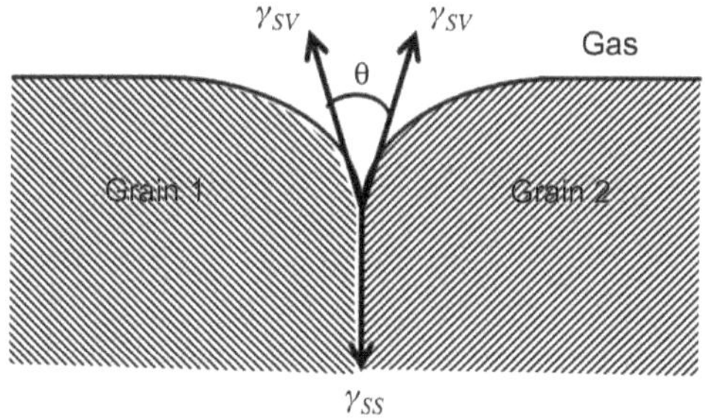

FIGURE 6.8 The force balance of the surface tension (γ_{SV}) and grain boundary surface tension (γ_{SS}) in a grain-boundary groove.

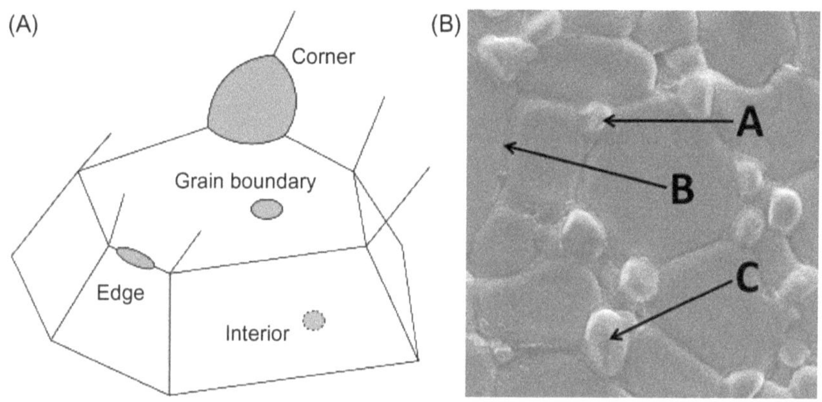

FIGURE 6.9 (A) Schematic illustration of the distribution of the second phases in a polycrystalline material located at a corner, an edge, a face, or the interior of a grain. (B) A two-dimensional distribution of the second phases in $CaCu_3Ti_4O_{12}$ (A: edge, B: grain boundary, and C: corner).

Concerning the equilibrium shapes of the second phases present at these four locations, it is important to realize how they are located in these positions during processing. For vapor phases, voids and pores will be considered. The former is related to the trapping of gas during solidification, usually located in the interior. The latter is relevant to the sintering of powder compacts, initially, they are located at edges and corners, but some or most pores may be trapped at grain boundaries or even inside the grains during grain growth. Regarding solid phases, they may develop through segregation or precipitation of impurities, usually located at grain boundary and interior. As for liquid phases, they may form due to the melting of the boundary solid phases, and if the wetting condition is satisfied, they may penetrate in edges and corners. The development of equilibrium shapes of the second phases will be discussed further in terms of their locations as follows.

1. The interior of a grain (or bulk). If the variation of surface free energy with the crystal orientation can be neglected, the second phases will

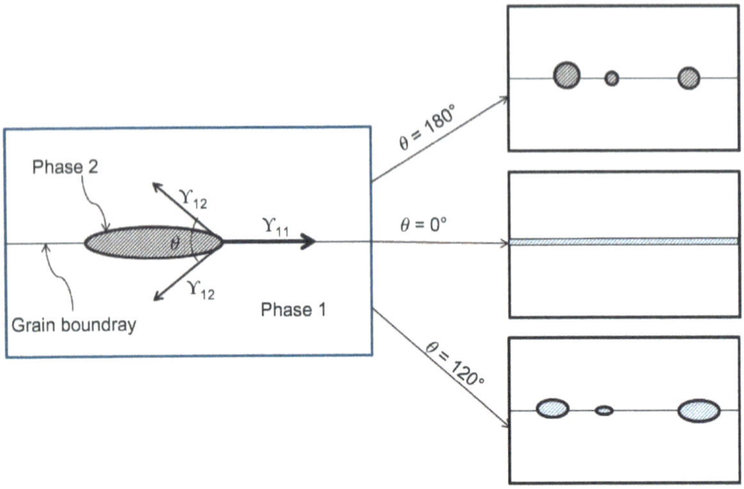

FIGURE 6.10 A surface tension force balance for a second phase 2, which can be vapors, liquids, or solids, intersects an grain boundary of phase 1, where the shapes of the second phases with three different dihedral angles at the grain boundary are schematically illustrated.

reveal a spherical shape; if not, i.e., surface free energy is anisotropic, the shape would be a polyhedron.

2. Face (or grain boundary). Consider the case of a second phase 2, which can be gases, liquids, or solids, located at grain boundary of phase 1. Take the cross-section of the boundary phase and boundary, as shown in Fig. 6.10.

At equilibrium, the formed angle θ is called the dihedral angle. Since the horizontal components of the surface free energy must balance,

$$\gamma_{11} = 2\gamma_{12}\cos\frac{\theta}{2} \tag{6.36}$$

and therefore,

$$\frac{\gamma_{12}}{\gamma_{11}} = \frac{1}{2\cos\frac{\theta}{2}} \tag{6.37}$$

Since the interfacial energy of either phase may be influenced by the crystal orientation and impurity, the dihedral angle may vary even for the same material. In addition, it should be noticed that Eq. (6.37) may be influenced by the inside pressure for gas phases and anisotropy for solid phases. Thus, liquid phases will be the main concerned issue in the following.

As per Eq. (6.37), two extremes will be taken into account in analogy to the wetting behavior (Section 6.5.1):

a. Nonwetting condition, $\gamma_{12} \gg > \gamma_{11}$, θ approximates to 180 degree. Phase 2 approaches a spherical shape, shown in Fig. 6.10.

b. Complete wetting condition, $\gamma_{12} \leqq 0.5\gamma_{11}$, θ approximates to 0 degree. Phase 2 tends to spread as a thin layer along the grain boundaries, shown in Fig. 6.10.

Regarding the dihedral angles between the two extreme cases, θ has a value in the vicinity of 120 degree, i.e., $\gamma_{12} \sim \gamma_{11}$, the shape of the phase 2 assumes a lenticular shape in general (Fig. 6.10).

3. Edge and corner. The second phase developed already at the edge and corner is the formation of pores during the sintering process. However, the morphological evolution of the channel pores is mostly determined by Rayleigh instability. Once the isolated pores are formed, the equilibrium shape should take into account the balance of the surface stresses of the phase boundaries. Concerning the second phases located at the edge and the corner, they may stem from the penetration of the liquid boundary phases. As far as the edge is concerned, it can be further inferred that the dihedral angle is in the range between 0 degree and 60 degree, where the liquid forms a continuous skeleton along the edge. Between 60 degree and 120 degree, the liquid no longer forms a continuous network, whose shape is suggested as triangular prisms at the three-grains junction on the basis of an equation analogy to Eq. (6.36), which basically is not sufficient to determine the equilibrium shape of the edge phase since the dimension along the edge is not taken into consideration. In addition, kinetically, Rayleigh instability may play a crucial role in the evolution of channel edge phase, which in turn would affect the shape. As far as the equilibrium shape is concerned, the thermodynamic viewpoint with respect to the balance of the surface stresses of the phase boundaries should be considered. In the light of the connection of the edge and corner, where the latter is the terminal of the former, the equilibrium shapes of both should be interrelated. Here, we will take a second phase 2 locating at the corner with the tetrahedral shape between three grains of phase 1, as shown in Fig. 6.11, where one of the tips is selected to analyze balance of the surface stresses, and the respective angles are as defined in the graph.

It is further assumed that the surface stresses of γ_{112} are equal, so the angles labeled θ are equal. Furthermore, underlying the geometry shown in Fig. 6.11, the functional relationships between dihedral angle φ and angles between edges must hold:

$$\cos\frac{\theta}{2} = \frac{1}{2\sin(\varphi/2)} \tag{6.38}$$

$$\cos(180 - \omega) = \frac{1}{\sqrt{3}\tan(\varphi/2)} \tag{6.39}$$

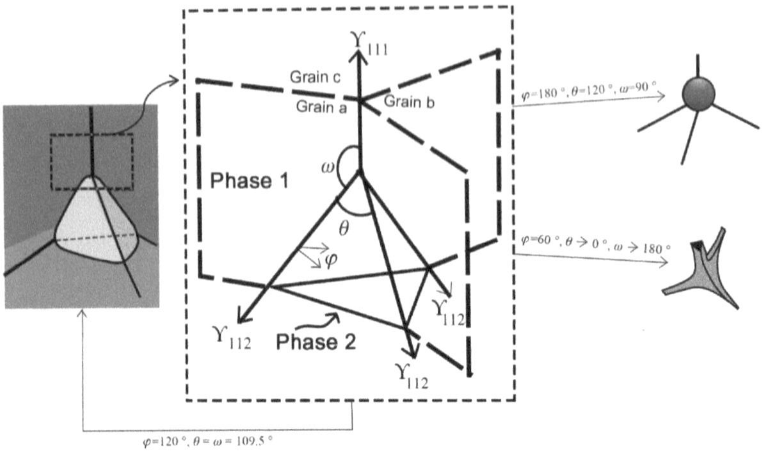

FIGURE 6.11 Schematic illustration of the surface tension force balance at the tip of phase 2 located at the grain corner of three grains of phase 1, showing the variation of the shapes of the second phase with three different dihedral angles at the grain edge.

Underlying the above two equations, the equilibrium shape of the second phase can be deduced. When $\varphi = 180$ degree, $\theta = 120$ degree, and $\omega = 90$ degree, phase 2 becomes spherical, as $\varphi = 120$ degree and $\theta = \omega = 109.5$ degree, the shape of phase 2 is like the tetrahedral SiO_4 units, and as $\varphi = 60$ degree and ω approaches 180 degree and θ to 0 degree, phase 2 will penetrate along the edges of the grains and in principle, a network-like structure would occur, which is represented in Fig. 6.11. As for $\varphi = 0$ degree, phase 2 not only penetrates along the edges of the grains but also spreads along the grain boundaries, which is consistent with the deduction of Eq. (6.36).

Surface or interfacial free energy plays a very important role in the essential driving force for many types of morphological evolution such as particle coarsening (Ostwald ripening), surface smoothing, and sintering. Underlying the thermodynamic viewpoint, the related phenomena are to reduce the total excess free energy associated with the interface itself. However, the fundamental basically is relevant to capillary force due to the pressure difference across the interface caused by its local mean curvature.

6.6 EFFECT OF INTERFACE CURVATURE ON THE EQUILIBRIUM PRESSURE BETWEEN TWO PHASES

The phenomenon concerning the existence of surface tension resulting in the equilibrium vapor pressure of a spherical droplet that is a function of the radius of curvature of the droplet was first discussed by Kelvin in 1871. Here, this viewpoint will be extended to the two-phase system (DeHoff, 1993).

Take into account a system composing of θ and δ phases, each containing n components and divided by curved interfaces. Moreover, the curvature is defined positive when the surface is convex relative to the δ phase, and thus the volume enclosed by the surface is that of the δ phase. The general expression of the entropy change due to the area change of the curved surface being tacitly applied to the system is:

$$dS_{sys} = dS^{\theta} + dS^{\delta} + S^{s}dA \tag{6.40}$$

where S^s is the specific superficial excess entropy for the surface. The entropy change of θ and δ phases can be written as

$$dS^k = \frac{1}{T^k}dE^k + \frac{P^k}{T^k}dV^k - \sum_{i=1}^{n}\frac{\mu_i^k}{T^k}dn_i^k \tag{6.41}$$

where k represents θ or δ. In addition, the isolation constraints should also be applied. The constant internal energy constraint is

$$dE_{sys} = 0 = dE^{\theta} + dE^{\delta} + E^{s}dA$$

$$dE^{\delta} = -(dE^{\theta} + E^{s}dA) \tag{6.42}$$

The constant volume constraint is:

$$dV_{sys} = 0 = dV^{\theta} + dV^{\delta} + V^{s}dA$$

The superficial excess volume V^s can be neglected and is assumed to be zero. Thus

$$-dV^{\theta} = dV^{\delta} \tag{6.43}$$

The total number of moles of each component is constant.

$$dn_{i,sys} = 0 = dn_i^{\theta} + dn_i^{\delta} + N_i dA \tag{6.44}$$

N_i is the specific interfacial excess of component i, defined as

$$N_i = n_i^{S}/A$$

$$dn_i^{\delta} = -(dn_i^{\theta} + N_i dA) \tag{6.45}$$

Substituting Eqs. (6.42), (6.43) and (6.45) into Eq. (6.41), and eliminating the dependent variables in an isolated system and collecting like terms gives

$$\begin{aligned} dS_{sys} = {} & \left(\frac{1}{T^{\theta}} - \frac{1}{T^{\delta}}\right)dE^{\theta} + \left(\frac{P^{\theta}}{T^{\theta}} - \frac{P^{\delta}}{T^{\delta}}\right)dV^{\theta} - \sum_{i=1}^{n}\left(\frac{\mu_i^{\theta}}{T^{\theta}} - \frac{\mu_i^{\delta}}{T^{\delta}}\right)dn_i^{\theta} \\ & + \left(S^{s} - \frac{1}{T^{\delta}}E^{s} + \sum_{i=1}^{n}\frac{\mu_i^{\delta}}{T^{\delta}}N_i\right)dA \end{aligned} \tag{6.46}$$

It has been proved that the change in surface area of a curved surface as it migrates is relevant to the volume swept through and to the local mean curvature, given by

$$dA = 2H\,dV \tag{6.47}$$

where mean curvature, H, is defined as $H = 1/2(\kappa_1 + \kappa_2)$ in which κ_1 and κ_2 are the principal curvature. Note that alternative definition for fluid mechanics, $H = (\kappa_1 + \kappa_2)$. Underlying Eq. (6.47),

$$dA = 2HdV^{\delta} = -2HdV^{\theta} \tag{6.48}$$

Combine Eq. (6.48) with Eq. (6.46) to eliminate dA

$$dS_{sys} = \left(\frac{1}{T^{\theta}} - \frac{1}{T^{\delta}}\right)dE^{\theta} + \left[\left(\frac{P^{\theta}}{T^{\theta}} - \frac{P^{\delta}}{T^{\delta}}\right) - \left(S^{s} - \frac{1}{T^{\delta}}E^{s} + \sum_{i=1}^{n}\frac{\mu_i^{\delta}}{T^{\delta}}N_i\right) \times 2H\right] \times dV^{\theta} - \sum_{i=1}^{n}\left(\frac{\mu_i^{\theta}}{T^{\theta}} - \frac{\mu_i^{\delta}}{T^{\delta}}\right)dn_i^{\theta} \tag{6.49}$$

Under the equilibrium conditions, dS_{sys} is assumed to be zero, so the coefficients of the differentials of Eq. (6.49) are simultaneously zero. The coefficient of dE leads to the condition for thermal equilibrium:

$$T^{\theta} = T^{\delta} \tag{6.50}$$

The coefficient of dn yields the condition for chemical equilibrium:

$$\mu_i^{\theta} = \mu_i^{\delta} \tag{6.51}$$

The coefficient of dV yields the condition for mechanical equilibrium:

$$\left(\frac{P^{\theta}}{T^{\theta}} - \frac{P^{\delta}}{T^{\delta}}\right) - \left(S^{s} - \frac{1}{T^{\delta}}E^{s} + \sum_{i=1}^{n}\frac{\mu_i^{\delta}}{T^{\delta}}N_i\right) \times 2H = 0 \tag{6.52}$$

$$P^{\theta} - P^{\delta} + \left(E^{s} - TS^{s} - \sum_{i=1}^{n}\mu_i N_i\right) \times 2H = 0 \tag{6.53}$$

$$\text{Let } \gamma = E^{s} - TS^{s} - \sum_{i=1}^{n}\mu_i N_i \tag{6.54}$$

Eq. (6.53) may be written as

$$P^{\delta} - P^{\theta} = 2\gamma H = \gamma\left(\frac{1}{r_1} + \frac{1}{r_2}\right) \tag{6.55}$$

where r_1 and r_2 are the principal radii of curvature.

For a spherical surface,

$$P^{\delta} - P^{\theta} = \gamma(2/r) \tag{6.56}$$

Eqs. (6.55) and (6.56) are the forms of the Gibbs–Thomson equation.

The defined γ of Eq. (6.54) can be further described as the specific interfacial free energy, which will be discussed in the following. By virtue of the analogous derivation for entropy S, and replacing entropy S (Eq. (6.40)) by E with $E = TdS - PdV + \sum_{i=1}^{n} \mu_i dn_i$, we can reach the final result:

$$dE_{sys} = TdS_{sys} - P^{\theta}dV^{\theta} - P^{\delta}dV^{\delta} + \sum_{i=1}^{n} \mu_i dn_{sys} + \left(E^s - TS^s - \sum_{i=1}^{n} \mu_i N_i\right)dA$$

Based on this equation, and further define $\gamma = \left(\frac{dE}{dA}\right)_{S,V^{\theta},V^{\delta},n_i}$, Eq. (6.54) can be achieved. Thus γ can be explained in terms of the change in internal energy of a two-phase system with the change of the internal area. Moreover, for a unary system, i.e., $n = 1$, it is possible to select the location of the dividing surface so that N_i is zero. Thus Eq. (6.54) becomes $\gamma = E^s - TS^s = F^s = G^s$. This is the reason for γ to be called the specific interfacial free energy. The Gibbs and Helmholtz free energy are identical because V^s is zero.

6.7 EFFECT OF THE INTERFACE CURVATURE ON THE EQUILIBRIUM SOLUBILITY BETWEEN TWO PHASES

While many books addressed this issue, (DeHoff, 1993) gave deep insight into understanding mechanistic aspect of the related process, which will be described as follows. Consider a system contains two phases, θ and δ with two components, A and B. Let δ phase be particles with convex surface, i.e., H is positive, dispersed in the θ phase. With regard to the component A in the θ phase, the variation of the chemical potential is a function of the temperature, pressure, and composition of the θ phase:

$$d\mu_A^{\theta} = \overline{V_A^{\theta}}dP^{\theta} - \overline{S_A^{\theta}}dT^{\theta} + \mu_{AB}^{\theta}dN_B^{\theta} \tag{6.57}$$

where $\overline{V_A^{\theta}}$ and $\overline{S_A^{\theta}}$ are partial molar entropy and volume for component A, respectively. Notice that component B is taken into account as the independent composition. The coefficient

$$\mu_{AB}^{\theta} = \left(\frac{d\mu_A}{dN_B}\right)_{T,P}^{\theta} \tag{6.58}$$

represents the change in chemical potential of component A with component B in the θ phase. Analogous expressions as Eq. (6.57) can be written for component B in θ phase and A and B in δ phase, and thus Eq. (6.58) is also applicable to the other three chemical potentials μ_{BB}^{θ}, μ_{AB}^{δ}, and μ_{BB}^{δ} in the system.

As mentioned in Section 6.6, the following conditions for a binary system, underlying Eqs. (6.50), (6.51) and (6.56), should be satisfied at equilibrium:

$$\mu_A^\theta = \mu_A^\delta$$
$$\mu_B^\theta = \mu_B^\delta$$
$$T^\theta = T^\delta$$

$$P^\delta - P^\theta = 2\gamma H$$

As the surface curvature of the δ phase is changed, the changes of the above equilibrium conditions between two phases require that

$$d\mu_A^\theta = d\mu_A^\delta \tag{6.59}$$

$$d\mu_B^\theta = d\mu_B^\delta \tag{6.60}$$

$$dT^\theta = dT^\delta = dT \tag{6.61}$$

$$dP^\delta - dP^\theta = 2\gamma dH \tag{6.62}$$

Substituting Eq. (6.57) and its analogous expression for component A in δ phase into Eq. (6.59) and combining Eqs. (6.61) and (6.62) gives

$$(\overline{V_A^\theta} - \overline{V_A^\delta})dP^\theta - (\overline{S_A^\theta} - \overline{S_A^\delta})dT - 2\gamma V_A^\delta dH + \mu_{AB}^\theta dN_B^\theta - \mu_{AB}^\delta dN_B^\delta = 0 \tag{6.63}$$

Substituting analogous expressions of Eq. (6.57) related to component B in θ and δ phases into Eq. (6.60) and combining Eqs. (6.61) and (6.62) gives

$$(\overline{V_B^\theta} - \overline{V_B^\delta})dP^\theta - (\overline{S_B^\theta} - \overline{S_B^\delta})dT - 2\gamma V_B^\delta dH + \mu_{BB}^\theta dN_B^\theta - \mu_{BB}^\delta dN_B^\delta = 0 \tag{6.64}$$

Let $dT = 0$ and $dP^\theta = 0$, Eqs. (6.63) and (6.64) become Eqs. (6.65) and (6.66), respectively:

$$-2\gamma\overline{V_A^\delta}dH + \mu_{AB}^\theta dN_B^\theta - \mu_{AB}^\delta dN_B^\delta = 0 \tag{6.65}$$

$$-2\gamma\overline{V_B^\delta}dH + \mu_{BB}^\theta dN_B^\theta - \mu_{BB}^\delta dN_B^\delta = 0 \tag{6.66}$$

By using the Gibbs–Duhem equation and Eq. (6.57) fixed at constant T and P, we can obtain

$$\mu_{BB} = -\frac{N_A}{N_B}\mu_{AB} \tag{6.67}$$

Eq. (6.67) applies separately to the θ or δ phase. With these relationships, Eqs. (6.65) and (6.66) can be rewritten

$$\mu_{AB}^\theta\left(\frac{dN_B^\theta}{dH}\right) - \mu_{AB}^\delta\left(\frac{dN_B^\delta}{dH}\right) = 2\gamma\overline{V_A^\delta} \tag{6.68}$$

$$-\frac{N_A^\theta}{N_B^\theta}\mu_{AB}^\theta\left(\frac{dN_B^\theta}{dH}\right)+\frac{N_A^\delta}{N_B^\delta}\mu_{AB}^\delta\left(\frac{dN_B^\delta}{dH}\right)=2\gamma\overline{V_B^\delta} \tag{6.69}$$

Eqs. (6.68) and (6.69) can be solved to give

$$\left(\frac{dN_B^\theta}{dH}\right)=2\gamma\left(N_A^\delta\overline{V_A^\delta}+N_B^\delta\overline{V_B^\delta}\right)\cdot\frac{N_B^\theta}{\mu_{AB}^\theta(N_A^\delta N_B^\theta-N_A^\theta N_B^\delta)} \tag{6.70}$$

$$\left(\frac{dN_B^\delta}{dH}\right)=2\gamma\left(N_A^\theta\overline{V_A^\delta}+N_B^\theta\overline{V_B^\delta}\right)\cdot\frac{N_B^\delta}{\mu_{AB}^\delta(N_A^\delta N_B^\theta-N_A^\theta N_B^\delta)} \tag{6.71}$$

It is noticed that in Eq. (6.70), $N_A^\delta\overline{V_A^\delta}+N_B^\delta\overline{V_B^\delta}=V^\delta$, where V^δ is the molar volume of the δ phase. It is further assumed that the θ phase is dilute in component B and thus the component A obeys the Raoult's law, yielding $\mu_{AB}^\theta\cong-RT$. And also assume that the δ phase is dilute in component A whose chemical potential can be described by the Henry's law, giving $\mu_{AB}^\delta=-\frac{RT}{N_A^\delta}$. Furthermore, since θ and δ phases are dilute in component B and A, respectively, $N_B^\theta\simeq0; N_A^\theta\simeq1; N_B^\delta\simeq1; N_A^\delta\simeq0; (N_A^\delta N_B^\theta-N_A^\theta N_B^\delta)\simeq-1$. Underlying these conditions, Eqs. (6.70) and (6.71) can be approximated as

$$\left(\frac{dN_B^\theta}{dH}\right)=\frac{2\gamma V^\delta}{RT}N_B^\theta \tag{6.72}$$

$$\left(\frac{dN_B^\delta}{dH}\right)=\frac{2\gamma\overline{V_A^\delta}}{RT}N_A^\delta \tag{6.73}$$

Eq. (6.72) is concerned about the variation of the concentration or solubility of component B in the θ phase with the surface curvature of the δ phase. Integrating Eq. (6.72) by separating the variables, gives

$$\int_{N_B^\theta(H=0)}^{N_B^\theta(H)}\frac{dN_B^\theta}{N_B^\theta}=\int_{H=0}^{H}\frac{2\gamma V^\delta}{RT}dH \tag{6.74}$$

$$\ln\left(\frac{N_B^\theta(H)}{N_B^\theta(H=0)}\right)=\frac{2\gamma V^\delta}{RT}H \tag{6.75}$$

Since H = 1/r for a spherical surface, Eq. (6.75) can further be rewritten as

$$N_B^\theta(H)=N_B^\theta(H=0)\exp\left(\frac{\gamma V^\delta}{kT}\frac{2}{r}\right) \tag{6.76}$$

Since the exponential term is usually small, and $e^x\approx1+x$, Eq. (6.76) becomes

$$N_B^\theta(H)\approx N_B^\theta(H=0)\left(1+\frac{\gamma V^\delta}{kT}\frac{2}{r}\right) \tag{6.77}$$

Eq. (6.77) indicates that the concentration of the component B in the θ phase in equilibrium with the δ phase with mean curvature H is larger than the concentration in the analogous system where the surface of the δ phase is flat; in other words, the solubility is enhanced as the particle size of the δ phase is reduced. For a nonspherical surface, Eq. (6.77) can be extended as

$$N_B^\theta(H) = N_B^\theta(H=0)\exp\left(\frac{\gamma V^\delta}{kT}\left(\frac{1}{r_1}+\frac{1}{r_2}\right)\right) \tag{6.78}$$

A similar result can be achieved for the component A in the δ phase by integrating Eq. (6.73) with $(N_A^\delta + N_B^\delta) = 1$:

$$\ln\left(\frac{N_A^\delta(H)}{N_A^\delta(H=0)}\right) = -\frac{2\gamma \overline{V_A^\delta}}{RT}H \tag{6.79}$$

We further obtain

$$N_A^\delta(H) \approx N_A^\delta(H=0)\left(1-\frac{\gamma \overline{V_A^\delta}}{kT}\frac{2}{r}\right) \tag{6.80}$$

Eq. (6.80) suggests that the equilibrium concentration of component A in the δ phase is decreased as the particle size is reduced.

Eqs. (6.77) and (6.78) are usually referred to the Thompson–Freundlich equation. According to these two equations, the concentration of solute in θ adjacent to small particles is higher by capillarity than that adjacent to the large particles. Thus the concentration gradient established in the θ phase produces a diffusion flux of solute from the region of high concentration around the small particles to that of lower concentration adjacent to the large particles. Consequently, the small particles would shrink but the large particles would coarsen. The above description is the fundamental for the kinetics of the particle coarsening process, usually called the Ostwald ripening process.

6.8 EQUILIBRIUM VACANCY CONCENTRATION CHANGES AT CURVED SURFACES: DRIVING FORCE FOR SINTERING

The issue concerning the equilibrium vacancy concentration change at curved surfaces is usually taken into account in the processes of diffusional creep and sintering. Regarding the modeling of the initial stage of sintering, twisted wires are usually used for the model experiment. After being sintered and cross-sectioned, the neck can be observed, as shown in Fig. 6.12. Take a small element MNOP of a curved surface of a neck shown in Fig. 6.12, where the force balance along the curve surface is presented.

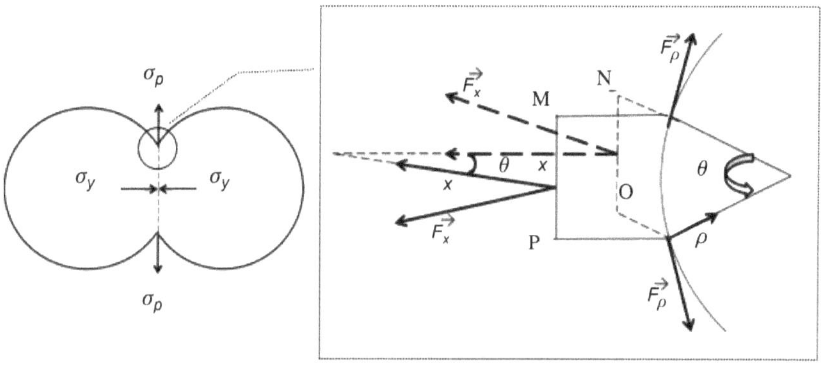

FIGURE 6.12 Schematic illustration of the surface tension force balance for an element at a curved surface of a neck in the initial stage of sintering, where the tensile stress would exert on the neck surface while a compressive stress exerts on the center of the contact.

The element is curved like a saddle so that the segments of the surface are described by two radii of curvature ρ and x, where each curvature assumes an angle of the same value θ. There are surface forces, $\overline{F}_x$ and $\overline{F}_\rho$ due to surface free energy, acting on element MNOP parallel to its surface.

$$\overline{F}_x = \gamma\,\overline{MP} = \gamma\overline{NO} \tag{6.81}$$

$$\overline{F}_\rho = \gamma\,\overline{MN} = \gamma\overline{OP} \tag{6.82}$$

where γ is surface free energy. According to Fig. 6.12,

$$\overline{MP} = \rho \sin\theta;\ \overline{MN} = x\sin\theta \tag{6.83}$$

For small θ, $\sin\theta \approx \theta$. Combining the above equations, gives

$$\overline{F}_x = \gamma\rho\theta \tag{6.84}$$

$$\overline{F}_\rho = -\gamma x\theta \tag{6.85}$$

The resultant force $\overline{F}$ acting perpendicularly on the element MNOP is expressed as

$$\overline{F} = 2\left[\overline{F}_x \sin\left(\frac{\theta}{2}\right) + \overline{F}_\rho \sin\left(\frac{\theta}{2}\right)\right] \tag{6.86}$$

Since
$\sin\frac{\theta}{2} \approx \frac{\theta}{2}$, we obtain

$$\overline{F} = \gamma\theta^2(\rho - x) \tag{6.87}$$

The area of the element is approximated as

$$\text{Area MNOP} = x\rho\theta^2 \tag{6.88}$$

Thus the stress exerting on the element MNOP can be obtained by dividing Eq. (6.87) by Eq. (6.88):

$$\sigma = \gamma\left(\frac{1}{x} - \frac{1}{\rho}\right) \tag{6.89}$$

Because $x \gg \rho$,

$$\sigma \simeq -\frac{\gamma}{\rho} \tag{6.90}$$

Eq. (6.90) indicates the occurrence of tensile stresses in the neck, which in turn develops a compressive stress of equal magnitude at the center of the contact, also illustrated in Fig. 6.12. Therefore, the concentration of vacancies in the neck would differ from the concentration in unstressed regions. Namely, stresses change the work necessary for the formation of a vacancy. The contribution by the stresses to the work of formation of a vacancy of volume Ω is given by the equation:

$$W = \pm\sigma\Omega \tag{6.91}$$

where "+" stands for compressive stresses and "−" for tensile stresses. The unstressed equilibrium vacancy concentration has been obtained (Eq. (4.21)):

$$\frac{n}{N} = \exp\left(-\frac{\Delta g_v}{kT}\right)$$

where $\Delta g_v = \Delta h_v - T\Delta s_{\text{vib}}$ is the free energy change associated with the formation of a single vacancy. Moreover, at low pressure, Δh_v is equivalent to the internal energy change ΔE_v. Thus the internal energy change for forming a vacancy in stressed regions is given by $\Delta E_v' = \Delta E_v \pm \sigma\Omega$. Therefore, the equilibrium concentrations of unstressed and stressed regions should be varied. Let $n/N = [V]$, and $[V]_0$, $[V]_{\text{comp}}$, and $[V]_{\text{tens}}$ represent equilibrium concentrations at unstressed, compression, and tensile regions, respectively. The concentration of vacancies near the center of the grain contact is given by

$$\begin{aligned}[V]_{\text{comp}} &= \exp\left(\frac{\Delta s_{\text{vib}}}{k}\right)\exp\left(-\frac{\Delta E_v'}{kT}\right) \\ &= \exp\left(\frac{\Delta s_{\text{vib}}}{k}\right)\exp\left(-\frac{\Delta E_v + \sigma\Omega}{kT}\right) \\ &= [V]_0\exp\left(-\frac{\sigma\Omega}{kT}\right)\end{aligned} \tag{6.92}$$

Since $\frac{\sigma\Omega}{kT} \ll 1$

$$\exp\left(-\frac{\sigma\Omega}{kT}\right) \approx 1 - \frac{\sigma\Omega}{kT} \tag{6.93}$$

$$[V]_{\text{comp}} = [V]_0\left(1 - \frac{\sigma\Omega}{kT}\right) \tag{6.94}$$

A similar reasoning enables us to obtain an expression for the vacancy concentration near the neck surface, where tensile stresses occur, namely

$$[V]_{\text{tens}} = \exp\left(\frac{\Delta s_{\text{vib}}}{k}\right)\exp\left(-\frac{\Delta E_v'}{kT}\right)$$
$$= \exp\left(\frac{\Delta s_{\text{vib}}}{k}\right)\exp\left(-\frac{\Delta E_v - \sigma\Omega}{kT}\right) \quad (6.95)$$
$$= [V]_0 \exp\left(\frac{\sigma\Omega}{kT}\right)$$

Furthermore,

$$\exp\left(\frac{\sigma\Omega}{kT}\right) \approx 1 + \frac{\sigma\Omega}{kT} \quad (6.96)$$

$$[V]_{\text{tens}} = [V]_0\left(1 + \frac{\sigma\Omega}{kT}\right) \quad (6.97)$$

Combining Eqs. (6.94) and (6.97), we can obtain an expression for the maximum difference of vacancy concentration between the vicinity of the neck surface and the vicinity of the intergrain contact center, $\Delta_1[V]$:

$$\Delta_1[V] = [V]_{\text{tens}} - [V]_{\text{comp}} \approx \frac{2[V]_0\sigma\Omega}{kT}. \quad (6.98)$$

The maximum difference of the vacancy concentration between the vicinity of the neck surface and the (unstressed) bulk of the grains, $\Delta_2[V]$ can also be obtained:

$$\Delta_2[V] = [V]_{\text{tens}} - [V]_0 \approx \frac{[V]_0\sigma\Omega}{kT} \quad (6.99)$$

Underlying Eqs. (6.98) and (6.99) concerning the difference of the vacancy concentration between neck surface and grain boundary or bulk, the related sintering kinetics can be derived.

Appendix I

Spectroscopic Notation

Consider Hamiltonian for multi-electron atom:

$$H \approx H_0 + H_1 + H_2$$

where

$$H_0 = \sum_i \left[-\frac{\hbar^2}{2m}\nabla_i^2 - \frac{Ze^2}{4\pi\varepsilon_0 r_i} + V_i(r_i) \right]$$

$$H_1 = \sum_{i<j} \frac{e^2}{4\pi\varepsilon_0 r_{ij}} - \sum_i V_i(r_i)$$

$$H_2 = \sum_i \lambda_i(r_i) L_i \cdot S_i;$$

H_0 represents central field terms, H_1 is residual Coulomb interaction, and H_2 indicates spin–orbit interaction; V_i (r_i) is the radially-symmetric "single-electron potentials" or central potential, accommodating the "average effect" of other electrons; λ_i are Lagrange multipliers.

For light atoms, $H_1 >> H_2$, can treat H_2 as a perturbation on $H_0 + H_1$, known as LS (or Russell-Saunders) coupling. For heavy atoms (large Z) (or ionized light atoms), spin–orbit coupling is strong, i.e., $H_2 >> H_1$, thus the spin, s, of an electron prefers to couple to its own orbital angular momentum, l, to give j, called jj coupling, where electrons become relativistic, and spin–orbit interaction dominates.

Concerning LS coupling for a given configuration, it is concluded:

1. The term with the largest permitted value of S lies lowest in energy, which is attributed to the fact that maximizing S makes spin wavefunction as symmetric as possible, and in turn tends to make spatial wavefunction antisymmetric, and reduces Coulomb repulsion.
2. For a given S, the term with the largest permitted value of L lies lowest in energy, which arises from the maximization of L tending to keep the electrons apart.

3. For a term with several levels: if the subshell is less than half full, the smallest value of J lies lowest in energy but if the subshell is more than half full, the highest value of J lies lowest in energy.

In spectroscopic notation $^{2S+1}L_J$, the superscript "$2S+1$" called"Spin multiplicity," where S is the total spin quantum number for the atom; the latter "L" represents the total orbital angular momentum quantum number for the atom: S: $L=0$, P: $L=1$, D: $L=2$, and F: $L=3$, etc.; the subscript "J" is the total angular momentum quantum number for the atom, depending how L and S are coupled. The possible values of the corresponding S, L, and J quantum number for two electrons resulting from a Clebsch–Gordan Series are $S=s_1+s_2, s_1+s_2-1, \ldots, |s_1-s_2|$, $L=l_1+l_2, l_1+l_2-1, \ldots, |l_1-l_2|$, and $J=L+S, L+S-1, \ldots, |L-S|$, respectively.

Appendix II

Band Structures of Mott–Hubbard Materials and the Transition Metal Ionic Compounds, and Polaron

Band theory rests on the assumption that the interaction between neighboring atoms is so large that each electron is shared equally by all like nuclei and is capable of correctly classifying a large number of crystalline solids as metals or insulators. However, as the interactions between neighboring atoms become weak, each electron would remain localized at a discrete atomic position. Usually, outer s and p electrons, which are primarily responsible for the binding energy of crystals, interact strongly with neighboring atoms. Therefore these are well described by a collective-electron model. By contrast, outer 4f or 5f electrons are tightly bound to their atomic nuclei and are screened from the neighboring atoms by 5s, 5p or by 6s, 6p, core electrons. Therefore the interactions between neighboring atoms become weak, resulting in the localization of these electrons. Regarding d electrons, the situation is more complicated. According to band theory, the material is predicted to be metallic if the density of states at the Fermi level is nonvanishing, i.e., $\rho(E_F)\neq 0$. If $\rho(E_F) = 0$, the material must be an insulator or semiconductor. Therefore, if there are an odd number of electrons per unit cell, the material cannot be an insulator. The electron configuration of the 3-d transition metal reveals that the d orbitals in most of the transition metals are partially filled, which would be considered to be a metal based on the band theory. However, they are intermediate in character, which is manifested by the fact that whereas in some crystals they exhibit the properties of localized electrons, in others they exhibit the properties of collective electrons, and in a few both types of d electrons appear to coexist. The reason can further be interpreted by the correlations effects (e.g., electron–electron repulsions). It is noticed that any purely electronic phenomenon that cannot be explained within the independent-electron (Hartree–Fock) approximation is generally regarded as a correlation effect.

The origin of band gaps and characters of the valence and conduction electron states in 3d transition metal compounds have been proposed in the Mott–Hubbard theory (Mott, 1990), in which the charge fluctuations (an electron hops between two neighboring atoms) involving the d–d Coulomb and exchange interactions (U) are strongly suppressed because of the high energies involved. It has been suggested that the reason for stabilizing localized electron states is that short-range d–d Coulomb repulsion of electrons can prevent formation of band states. The Coulomb repulsion is described in terms of a correlation energy, called Hubbard-U, which is relevant to the energy for transferring an electron between two adjacent sites. Hubbard-U is equivalent to the difference between ionization energy and electron affinity energy. According to Mott–Hubbard models, as shown in Fig. A.1, if $U > W$ (band width), the d band of the transition metal is splitted into subbands. Thus an energy barrier U must be overcome as an electron transfers, resulting in insulating materials. However, as W approaches U, the bands overlap, where no energy gap is present. Thus beyond this point W_c, the material is metallic.

Regarding transition metal ionic compounds, another parameter would have a significant influence on the band structures, i.e., Δ gap, representing a gap between anion band and the upper d band of the metal (Zaanen et al., 1985). Nevertheless, the d band of the transition metal is still needed to take the Mott–Hubbard model into account. For $U < W$, Δ in general is much larger than U, and thus this compound behaves like d band metals, in which both holes and electrons are heavy, e.g., the high-temperature phases of V_2O_3, Ti_2O_3, TiO, and CrO_2. As for $U > W$, three cases can be distinguished. Concerning $\Delta < U$ and also $< W$, the band structure is like a semimetal with an anion band overlapping with the upper d band, called p-type metals with holes in the anion valence band (light holes), e.g., CuS, CuSe, and NiSe,

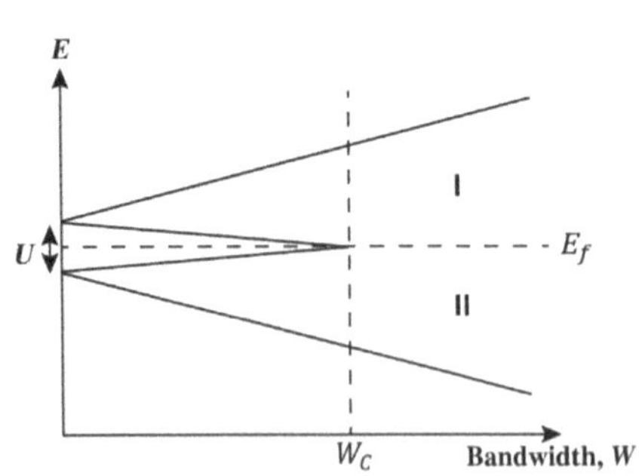

FIGURE A.1 Hubbard subbands as a function of band width, where W_c indicates the critical point for the metal–insulator transition; I and II represent upper and lower Hubbard band, respectively; E_f is the Fermi level.

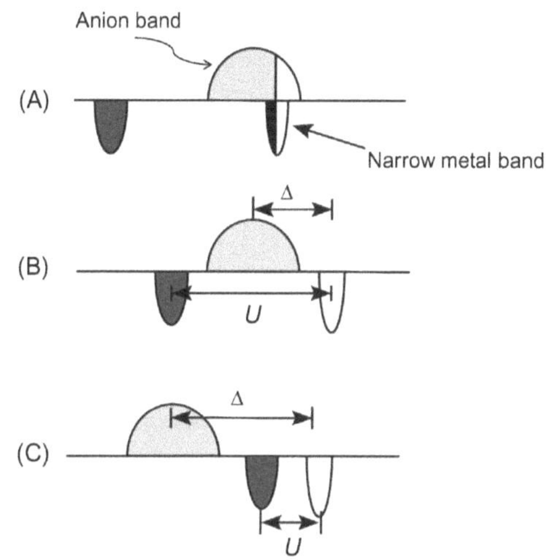

FIGURE A.2 Schematic illustrations of the band structures of the transition metal ionic compounds for $U > W$: (A) $\Delta < U$ and also $< W$, called p-type metals; (B) $\Delta < U$ but $> W$, called charge-transfer insulator (semiconductor); and (C) $\Delta > U$, behaving like Mott–Hubbard insulator.

shown in Fig. A.2A. For $\Delta < U$ but $> W$, this compound is called a charge-transfer insulator (semiconductor), where $E_g \sim \Delta$, directly related to the electronegativity of the anion and the Madelung potential, both of which tend to stabilize an ionic ground state and an inert-gas valence configuration of the anion, where holes are light (anion valence band) and electrons are heavy (d bands), e.g., $CuCl_2$, $CuBr_2$, CuO, $NiCl_2$, $NiBr_2$, and NiI_2, shown in Fig. A.2B. With respect to $\Delta > U$, this compound behaves like a Mott–Hubbard insulator, where $E_g \sim U$, and both holes and electrons move in d bands and are heavy, e.g., V_2O_3, Ti_2O_3, Cr_2O_3, and their halides, shown in Fig. A.2C.

One of the most important phenomena in transition metal oxides is concerned with the metal–insulator transitions, depending significantly on the defect concentration at constant temperature and pressure. As the defect concentration increases, the band structure would be changed, leading to the change of the conduction and magnetic behavior and even making charge-transfer and Mott–Hubbard insulators transform each other. In addition, it should be noticed that semimetal is different from the half-metal materials. The feature of the half-metal materials is that it is a solid with an unusual electronic structure. For electrons of one spin, it is a metal with a Fermi surface whose band is partially filled, but for the opposite spin there is a gap in the spin-polarized density of states, like a semiconductor or insulator. This results in conducting behavior for only electrons in the first spin orientation, e.g., magnetite and lanthanum strontium manganite (LSMO) with colossal magnetoresistance, described in Section 3.5.6.3. For conductive materials such as LSMO, their electrons responsible for the electronic conductivity can at the same time give rise to magnetism, called itinerant electrons.

As the electrons are localized, electron–phonon couplings would be strong. When a localized electron remains some time in the vicinity of a particular ion, the lattice will be deformed. The local change in the dielectric constant and the elastic relaxation of the system brings about a potential well where the electron can form a bound state called Landau trapping or self-trapping. In a 3d transition metal, the overlap life-time of an electron at an ionic position is of the order of 10^{-15} s. This short time does not make it possible that its own electrical field can polarize the crystal. However, if the electron moves in a very narrow band (may be constructed by the states of the impurity defects) or due to some other mechanism, the electron is near the core ion for $\sim 10^{-12}$ s. Then the electric field of the electron will attract the positive ions and push away the negative ions. The effective electric field at the electronic position decreases because of the increase of the dielectric constant due to both the polarization and the ionic displacement. If the electron tries to move away, it will feel a force that tends to restore it at the initial position. The electrostatic energy plus the elastic energy add to give the potential barrier (hopping energy) that should be overcome to release the charge from the potential well. This bound state associating charge with lattice distortion is called polaron (Tsuda et al., 1991).

When the electron–phonon interacts, the electron becomes heavier, and the distortion is inadequate to self-trap the carrier, but still serves to slow the carrier, wherein this state is called large polaron, moving in a band with a slightly enhanced mass. When the interaction becomes larger in a compound, electrons push away negative ions further, overlap between neighboring orbitals becomes smaller, and finally the electrons fall into localized states called the small polaron. The electronic wave function of the small polaron is localized in a radius of the order of a lattice parameter.

There exist two mechanisms for the small polaron to move from site to site in the lattice. At low temperatures, small polarons exhibit band-like properties and move with quantum tunneling; but they possess a large effective mass which impedes their mobility considerably. As the temperature increases, the band becomes narrow. The electronic behavior becomes thermally activated and indistinguishable from that of the localized hopping-type materials.

The self-trapping energy (or formation energy of polaron), E_{st}, depending on the Hubbard energy (U), temperature, and carrier concentration, may be considered as the energy which is required in order to produce the self-trapped (or polaronic) state from the lowest delocalized (band) state of the quasi-particle. The self-trapping energy can be decomposed into a localization ($\mathbf{E}_{loc}$) and a relaxation energy ($\mathbf{E}_{rel}$), i.e., $\mathbf{E}_{st} = \mathbf{E}_{loc} + \mathbf{E}_{rel}$. The localization energy $\mathbf{E}_{loc}$ corresponds to the energy cost for exciting a carrier from the delocalized band-like state into the respective molecular orbital states inside continuum of band energies (i.e., the kinetic energy gain from having a delocalized carrier in an undistorted lattice). $\mathbf{E}_{rel}$ (negative value) is the energy including the polarization and distortion energy (If Jahn-Teller distortion occurs, its energy should be taken into account.) gain upon lattice distortion in the presence of the localized carrier. Self-trapping and small polaron conduction are likely to occur if $\mathbf{E}_{loc}$ is small or if $\mathbf{E}_{rel}$ is large (in its absolute value), leading to negative $\mathbf{E}_{st}$. If the calculated value of E_{st} is negative, the localized state is considered to be stable with respect to the band of the delocalized states. If E_{st} is found to be positive, the localized state is certainly unstable and the electron and hole can exist only in the delocalized state. Small localization energy can be expected in cases where the orbital symmetry that hosts the polaron has a high density of states close to the band edge. A large $\mathbf{E}_{rel}$ is expected when the respective orbital symmetry interacts with the ligands forming a bonding or an antibonding state, whereas non-bonding symmetries induce generally smaller relaxation effects upon electron addition or removal. A high 3d density of states close to the conduction band or valence band extreme should be avoided, because it tends to cause large effective masses and to support self-trapping.

Bibliography

Abbaschian, R., Abbaschian, L., Reed-Hill, R.E., 1994. Physical Metallurgy Principles, fourth ed. Cengage Learning, Boston, MA.

Allen, S.M., Thomas, E.L., 1999. The Structure of Materials. John Wiley & Sons, New York.

Bacskay, G.B., Reimers, J.R., Nordholm, S., 1997. The mechanism of covalent bonding. J. Chem. Educ. 74 (12).

Baldinia, M., Muramatsub, T., Sherafatic, M., Maob, H.K., Malavasie, L., Postorinog, P., et al., 2015. Origin of colossal magnetoresistance in $LaMnO_3$ manganite. Proc. Natl. Acad. Sci. U.S.A 112 (35), 10869.

Banerjea, D., 1993. Coordination Chemistry. Tata McGraw Hill Publishing Company Limited, New Delhi.

Barrett, C.R., Nix, W.D., Tetelman, A.S., 1973. The Principles of Engineering Materials. Prentice Hall, New Jersey.

Barsoum, M.W., 1997. Fundamentals of Ceramics. McGraw Hill, New York.

Benedek, N.A., 2014. Origin of ferroelectricity in a family of polar oxides: The Dion-Jacobson phases. Inorg. Chem. 53, 3769.

Brown, I.D., 2009. Recent developments in the methods and applications of the bond valence model. Chem. Rev. 109, 6858.

Brown, I.D., Altermatt, D., 1985. Bond-valence parameters obtained from a systematic analysis of the inorganic crystal structure database. Acta Cryst B41, 244.

Brown, I.D., Shannon, R.D., 1973. Empirical bond-strength-bond-length curves for oxides. Acta Cryst A29, 266.

Bunget, I., Popescu, M., 1984. Physics of Solid Dielectrics. Elsevier, Amsterdam.

Burdett, J.K., McLarnan, T.J., 1984. An orbital interpretation of Pauling's rules. Am. Mineral. 69, 601421.

Cahn, R.W., Haasen, P., 1983. Physical Metallurgy, third ed., revised and enlarged edition, Elsevier Science Publishers B.V, Amsterdam.

Cammarata, R.C., Sieradzki, K., 1994. Surface and interface stresses. Annu. Rev. Mater. Res. 24, 215.

Catalan, G., Scott, J.F., 2009. Physics and applications of bismuth ferrite. Adv. Mater. 21, 2463.

Chiang, Y.M., Birnie, D.P., Kingery, W.D., 1997. Physical Ceramics Principles for Ceramic Science and Engineering. John Wiley, New York.

Choi, G.M., Tuller, H.L., 1988. Defect structure and electrical properties of single-crystal $Ba_{0.03}Sr_{0.97}TiO_3$. J. Am. Ceram. Soc. 71 (4), 201.

Cohen, R.E., 1992. Origin of ferroelectricity in perovskite oxides. Nature 358, 136.

Cottrell, A.H., 1953. Dislocations and Plastic flow in Crystals. Oxford University Press, London.

Cullity, B.D., 1972. Introduction to Magnetic Materials. Addison-Wesley, Reading, MA.

Dagotto, E., Hotta, T., Moreo, A., 2001. Colossal magnetoresistant materials: the key role of phase separation. Phys. Rep. 344 (1), 153.

DeHoff, R.T., 1993. Thermodynamics in Materials Science, International editions. McGraw-Hill Inc, New York.

Dunitz, J.D., Orgel, L.E., 1960. Stereochemistry of ionic solids. Adv. Inorg. Radiochem 2, 1.

Fang, T.T., Chiu, T.Y., 2011. Polarization dynamics of polar nano-regions in $Sr_{0.5}Ba_{0.5}Nb_2O_6$ doped with combinations of Ce and Cr. Acta Mater. 59, 1692.

Fang, T.T., Chung, H.Y., 2011. "Reassessment of impedance spectra and dielectric responses of undoped and $CaSiO_3$-doped $CaCu_3Ti_4O_{12}$". J. Appl. Phys. 109, 014102.

Fang, T.T., Chung, H.Y., 2011. Significant effects of electron transport on the electrochemical properties of Ni^{2+}- and Al^{3+}-doped $LiMn_2O_4$. Electrochem. Solid-State Lett. 14, A49.

Fang, T.T., Wang, H.Y., 2011. Reassessment of copper and titanium valences and excess holes in oxygen 2p levels of $CaCu_3Ti_4O_{12}$. J. Electrochem. Soc. 158 (9), G207.

Fang, T.T., Chung, H.Y., Lee, C.H., 2010. Defects, structure changes, and the effect of random fields on the displacement of off-center ions in $Sr_{0.5}Ba_{0.5}Nb_2O_6$ doped with combinations of Ce and Cr. J. Am. Ceram. Soc 93, 2339.

Fang, T.T., Wang, H.Y., Kuo, J.C., 2011. Role of strained nano-regions in the formation of sub-grains in $CaCu_3Ti_4O_{12}$. J. Appl. Phys. 110, 024103.

Freysoldt, C., Grabowski, B., Hickel, T., Neugebauer, J., Kresse, G., Janotti, A., et al., 2014. First-principles calculations for point defects in solids. Rev. Mod. Phys. 86 (1), 253.

Gallina, V., Omini, M., 1964. Vacancy as a phonon field perturbation (part IV). Phys. Status Solidi 7, 405.

Giacovazzo, C., Monaco, H.L., Viterbo, D., Scordari, F., Gilli, G., Zanotti, G., et al., 1992. Fundamentals of Crystallography. International Union of Crystallography Oxford University Press, New York.

Guy, A.G., Hren, J.J., 1973. Elements of Physical Metallurgy. Addison-Wesley, Reading, MA.

Hench, L.L., West, J.K., 1990. Principles of Electronic Ceramics. John Wiley and Sons, New York.

Hirth, J.P., Lothe, J., 1968. Theory of Dislocations. McGraw Hill, New York.

von Hippel, A.R., 1954. Dielectrics and Waves. John Wiley & Sons, New York.

Howe, J.M., 1997. Interfaces in Materials. John Wiley & Sons, Inc, New York.

Hull, D., Bacon, D.J., 1984. Introduction to Dislocations, third ed Pergamon Press, Oxford.

Jackson, M., Moskowitz, B., Bowles, J., 2011. Interpretation of low-temperature data part III: the magnetite Verwey transition. IRM Quar 20 (4), 1.

Jaffe, B., Cook Jr., W.R., Jaffe, H., 1971. Piezoelectric Ceramics. Academic Press, London.

Kim, Y.-S., Park C. H., 2009. Rich variety of defects in ZnO via an attractive interaction between O vacancies and Zn interstitials: origin of n-type doping. Phys. Rev. Lett.102, 086403.

Kingery, W.D., Bowen, H.K., Uhlmann, D.R., 1976. Introduction to Ceramics, second ed. John Wiley and Sons, New York.

Kittel, C., 1986. Introduction to Solid State Physics, sixth ed. John Wiley and Sons, New York.

Kofstad, P., 1983. Nonstoichiometry, Diffusion, and Electrical Conductivity in Binary Metal Oxides. Krieger Publications, Malabar, FL.

Kröger, F.A., 1974. The Chemistry of Imperfect Crystals, second ed, vol. 2. North-Holland Publishing Company, Amsterdam.

Lany, S., Osorio-Guillén, J., Zunger, A., 2007. Origins of the doping asymmetry in oxides: hole doping in NiO versus electron doping in ZnO. Phys. Rev. B 75, 241203R.

Miura, K., Azuma, M., Funakubo, H., 2011. Electronic and structural properties of ABO_3: role of the B-O coulomb repulsions for ferroelectricity. Materials 4, 260.

Moffatt, W.G., Pearsall, G.W., Wulff, J., 1971. Structure and Properties of Materials. Wiley, New York.

Mott, N.F., 1990. Metal-Insulator Transitions, second ed. Taylor & Francis, New York.

Moulson, A.J., Herbert, J.M., 1990. Electronics, Materials, Properties, and Applications. Chapman and Hall, London.

Nye, J.F., 1985. Physical Properties of Crystals. Oxford University Press, New York.

Pouchard, M., Doumerc, J.P., Villesuzanne, A., 2008. Status of trivalent copper and charge-transfer excitons in high-TC cuprates. Inorg. Chem. 47, 8487.

Ramirez, A.P., 1997. Colossal magnetoresistance. J. Phys. Condens. Matter 9, 8171.

Read Jr., W.T., 1953. Dislocations in Crystals. McGraw-Hill, New York.

Rioux, F., 2001. The covalent bond in H_2. Chem. Educ. 6, 288.

Rodriguez-Carvajal, J., Rousse, G., Masquelier, C., Hervieu, M., 1998. Electronic crystallization in a lithium battery material columnar ordering of electrons and holes in the spinel $LiMn_2O_4$. Phys. Rev. Lett. 81, 4660.

Schwarz, W.H.E., 2010. The full story of the electron configurations of the transition elements. J. Chem. Educ. 87 (4), 444.

Senn, M.S., Wright, J.P., Attfield, J.P., 2012. Charge order and three-site distortions in the Verwey structure of magnetite. Nature 481, 173.

Smart, L., Moore, E., 1992. Solid State Chemistry. Chapman & Hall, London.

Smith, C.S., 1952. Grain shapes and other metallurgical applications of topology. Metal Interfaces. American Society for Metals, Metals Park, OH, Also, see discussion by Neumann, von following article, pp. 108–110.

Smyth, D.M., 1988. In: Yan, M.F. (Ed.), Ceramic Superconductors, II, Research Update 1988. The American Ceramic Society, pp. 1–10.

Smyth, D.M., 2000. The Defect Chemistry of Metal Oxides. Oxford University Press, New York.

Swalin, R.A., 1961. Thermodynamics of Solids. Wiley, New York.

Tilley, R.J.D., 2004. Understanding Solids. John Wiley & Sons, New York.

Tsuda, N., Nasu, K., Yanase, A., Siratori, K., 1991. Electronic Conduction in Oxides. Springer-Verlag, New York.

Turton, R., 2000. The Physics of Solids. Oxford University Press, New York.

Verhoeven, J.D., 1975. Fundamentals of Physical Metallurgy. John Wiley & Sons, New York.

Walsh, A., Watson, G.W., 2005. The origin of the stereochemically active Pb(II) lone pair: DFT calculations on PbO and PbS. J. Solid State Chem. 178, 1422.

Wang, S.G., Qiu, Y.X., Fang, H., Schwarz, W.H.E., 2006. The challenge of the so-called electron configurations of the transition metals. Chem. Eur. J 12, 4101.

Wang, Z.L., Kang, Z.C., 1998. Functional and Smart Materials, Structural Evolution and Structure Analysis. Plenum Press, New York.

West, A.R., 1986. Solid State Chemistry and Its Applications. John Wiley & Sons, New York.

Yanagida, H., Koumoto, K., Miyayama, M., Yamada, H., 1996. The Chemistry of Ceramics. John Wiley & Sons, New York.

Yang, C.W., Fang, T.T., 2012. Structures and development mechanism of the anti-phase boundaries in Sr_2FeMoO_6. J. Electrochem. Soc. 159 (3), 35.

Zaanen, J., Sawatzky, G.A., Allen, J.W., 1985. Band gaps and electronic structure of transition-metal compounds. Phys. Rev. Lett. 55 (4), 418.

Zhang, J.X., He, Q., Trassin, M., Luo, W., Yi, D., Rossell, M.D., et al., 2011. Microscopic origin of the giant ferroelectric polarization in tetragonal-like $BiFeO_3$. Phys. Rev. Lett. 107, 147602.

Index

Note: Page numbers followed by "*f*" refer to figures.

A

Alpha silicon carbide (α-SiC), 62–63
Antiferroelectricity (AFE), 68
Antiphase boundaries (APBs), 71–73
Atoms with a single electron, 3–6
Atoms with more than one electron, 6–11
  energetic d-orbital collapse of free neutral atoms, 8–11
  penetration and shielding, 7–8
Attractive potential energy, 17
Aufbau principle, 9–10
Avogadro's number, 20

B

Band theory, 201–202
Barycenter principle, 49
Beta silicon carbide (β-SiC), 62–63
$BiFeO_3$, 69–70
  rhombohedral structure of, 70*f*
Binary metal oxides, point defects in, 92–94
Binding energy, 15, 30
Bond energy, defined, 13
Bond length, effects of the Jahn–Teller distortions on, 56–58
Bond strength, 43–44
Bond valence, 43–44
Bonding within crystal structures, 13
  covalent bonds, 29–31
    formation energy and mechanism, 30–31
  ionic crystals, bonding in, 13–29
    neutral atoms, energy for forming the ions from, 13–15
  metallic bonds, 31–32
    cohesive energy of metals, 31–32
  potential-well concept, effect of bonding on material properties by means of, 32–35
    elastic modulus, 32–33
    melting temperatures, 33
    thermal property, 33–35
  thermodynamic viewpoint of bond energy, 15–29
    Born–Haber cycle, 29
    electrostatic energy, 15–18
    forces between ions in ionic crystal, 18–22
    lattice energy, approximation of, 18–22
    repulsive energy, 18
    van der Waals attractive energy, 22–26
    zero-point energy, 26–28
Bond-length curve, 44
Born exponent, 18
Born's repulsive energy, 28
Born–Haber cycle, 28–29
Bravais lattice, 38, 38*f*, 64, 179
Brookite, 43
Brouwer diagram, 113–126
Burgers vector, 132–134, 133*f*, 136, 140–141, 161, 165, 171–172

C

$CaCu_3Ti_4O_{12}$, 77–79, 78*f*
Capillary forces, 188
Cation–cation Columbic repulsions, 46
Charge transfer insulator, 202–203
Climb motion, 135–136
  versus slip, 135–136
Coherent boundary, 174–175
Coherent phase boundary, 175*f*
  partially, 176*f*
Cohesive energy, 15
  of metals, 31–32
Coincidence site lattice, 173
Colossal magnetoresistance (CMR), 75–76
Conventional Madelung constant, 17
Coordination number (CN), 46, 59
Coulomb interaction, 15–16
Coulomb repulsion, 68, 202
Covalent bonding, 29–31

Covalent bonding (*Continued*)
formation energy and mechanism of, 30–31
Critical resolved shear stress (CRSS), 131, 137–138
Cross slip, 165
Crystal field
in octahedral coordination, 47–50, 47*f*, 48*f*
splitting, 49
in tetrahedral coordination, 50–51, 51*f*
Crystal field stabilization energy (CFSE), 51–53
Crystallographic slip, 132
Cubic closely packed (CCP) structures, 45
$CuCr_2O_4$, 59
$CuFe_2O_4$, 58–59
Curie temperature, 77

D

d electron configuration, effects of
on the radii of transition metal ions, 53–55
Debye density function, 28
Debye model, 87
Dipole
field produced by, 22–24
induced, 24–26
Dipole moment, 24
Dislocations
bowing of, 149–150
elementary geometric characteristics of, 132–137
edge dislocation, 135–136
mixed dislocation, 136–137
screw dislocation, 136
energy of, 145–147
extended dislocations, 159–167
FCC crystals, stacking faults in, 161–162
partial dislocations, 160–161
force between, 150–151
forces upon, 148–149
line tension of, 147–148
observations of, 131
plastic flow associated with the slip of, 138–142
reactions between, 154–159
dislocation nodes, 154–156
intersections of dislocations, 157–159
kinks and jogs, 156–157
stress fields of, 142–145
constitutive equations with Lame constants, 142–143
edge dislocation, 145
screw dislocation, 143–144
d-orbital collapse, 8–11
Doubly-charged oxygen vacancies, 119–120, 119*f*

E

e orbitals, 50
Edge dislocation, 135–136
stress field of, 145
(8-*N*) rule, 31
Einstein model, 87
Einstein–Bose statistics, 87
Elastic modulus, 32–33
Electron affinity (EA), 13–14
Electron configuration of atoms, 3
atoms with a single electron, 3–6
atoms with more than one electron, 6–11
energetic d-orbital collapse of free neutral atoms, 8–11
penetration and shielding, 7–8
Electron density distribution, 5–6
Electron ratio, 85
Electronegativity (EN), 13
Electron–electron repulsion, 7, 9–10, 201–202
Electron–nuclear attraction, 7
Electron–phonon couplings, 204
Electron–phonon interacts, 204
Electropositive atom, 14
Electrostatic bond strength (e.b.s.), 42
Electrostatic energy, 15–18
Electrostatic field produced by electrical dipole, 23*f*
Electrostatic valence rule, 42–44
Equilibrium at constant pressure, 86
Equilibrium concentration of point defects, 86–92
Equilibrium vacancy concentration, 194–197
Extended dislocations, 159–167
Extrinsic nonstoichiometry, 100–109
donors and acceptors, 100–102
issue concerning the choice of compensating defects, 106–107
metal-deficient oxides, 108–109
oxygen deficient oxides, 107
purely extrinsic electronic disorder/electronically compensated solutes, 104–105

purely extrinsic ionic disorder/ionically compensated solutes, 102–104

F

Face-centered-cubic (FCC) crystals, stacking faults in, 161–162
Ferroelectricity (FE), 68
Frank partial dislocation, 165
Frank's rule, 155–156
Free energy, 37, 177
Free neutral atoms, energetic d-orbital collapse of, 8–11
Free surface, 179–181
Frenkel defects, 111–112
Frenkel disorder, 94, 96
Fully-filled 3d subshell, 10

G

Gibbs free energy, 15, 86
Glide, defined, 132
Glissile dislocation, 165
Goldsmith tolerance factor, 69–70
Grain boundaries, 169–173
  coincidence site lattice, 173
  tilt boundary, 171–173
  twist boundary, 173

H

Half-metal materials, 203–204
Harmonic crystal, 33–35
HCP structure, 39, 40*f*
Heisenberg uncertainty principle, 26–27
Helmholtz free energy, 15, 86–87
Hess' law, 29
High-resolution transmission electronic microscope (HRTEM), 71, 72*f*, 73*f*
Hole-doped manganites, 75
Hooke's law, 20–21
Hopping energy, 204
Hubbard-U, 202
Hume–Rothery rules, 85
Hund's rule, 49–50, 56–58
  first rule, 9–10

I

Impurity atoms, factors affecting the solubility of
  in interstitial solid solution, 85–86
  in substitutional solid solution, 85
Impurity defects, 83
Incoherent boundary, 175
Induced dipole moment, 24–26
Interface curvature, effect of
  on equilibrium pressure between two phases, 188–191
  on equilibrium solubility between two phases, 191–194
Interfaces, defined, 169
Interfaces of phases, 181–188
  equilibrium shapes of grains, 182–183
  presence of second phases, 184–188
  wetting, 181–182
Interphase boundaries in solids, 174–176
  coherent boundary, 174–175
  incoherent boundary, 175
  partially coherent/semicoherent boundary, 175–176
Interstitial solid solution, 83, 85–86
Intrinsic nonstoichiometry, 97–100
  intrinsic electronic reaction, 100
  metal-deficient oxides, 98–99
  oxygen-deficient oxides, 97–98
Ionic and covalent structures, 59–80
  perovskite ABO_3, 67–80, 67*f*
    $CaCu_3Ti_4O_{12}$, 77–79, 78*f*
    effect of defect (anion deficiency), 69
    Jahn–Teller effects, 69
    $LaBa_2Cu_3O_7$, 76–77, 76*f*
    manganite, 75–76
    size effects, 67–68
    $SrBaNb_2O_6$, 79–80, 79*f*
    $YBa_2Cu_3O_{7-x}$, 76–77
  silicon carbide, 62–63, 62*f*
  spinel (AB_2O_4), 64–67, 65*f*
    $LiMn_2O_4$, 65
    magnetite, 65–67
  TiO_2, 63, 63*f*
  zinc oxide, 61–62, 61*f*
  ZrO_2, 63–64, 64*f*
Ionic bonds, 13–16
Ionic crystals
  bonding in, 13–29
    neutral atoms, energy for forming the ions from, 13–15
  forces between ions in, 18–22
  nonstoichiometric, 96–107
Ionic solids
  defect equilibrium in, 109–126
  point defects in, 92–127
Ionic structures, 40–46
  Pauling's fifth rule, 46

Ionic structures (*Continued*)
Pauling's first rule, 41–42
Pauling's fourth rule, 46
Pauling's second rule, 42–44
Pauling's third rule, 44–45
Ionic structures, structural distortion in, 46–59
crystal field stabilization energy (CFSE), 51–53
d electron configuration, effects of
on the radii of transition metal ions, 53–55
Jahn–Teller distortions on the bond length, 56–58
Jahn–Teller effect
in the octahedral field, 55–56
in the tetrahedral field, 58–59
octahedral coordination, crystal field in, 47–50, 47*f*, 48*f*
structure distortions arising from asymmetric electron density/inert pair effect, 59
tetrahedral coordination, crystal field in, 50–51, 51*f*
Ionization energy (IE), 13–14
Itinerant electrons, 203–204

J

Jahn–Teller distortions, 55–59
on the bond length, 56–58
Jahn–Teller effect, 65, 69
$BiFeO_3$, 69–70
in the octahedral field, 55–56
Sr_2FeMoO_6, 70–74
in the tetrahedral field, 58–59
Jogs, 156–157

K

Kinks, 156–157
Kröger–Vink diagram, 113–126

L

$LaBa_2Cu_3O_7$, 76–77, 76*f*
Lame constants, 142
Landau trapping, 204
Lanthanum strontium manganite (LSMO), 203–204
Lattice constant, 38–39, 160
Lattice energy, approximation of, 18–22
Lattice parameter, 38–39
Lattice translation vectors, 38–39
$LiMn_2O_4$, 65
Line defects in crystalline solids, 129
critical resolved shear stress (CRSS), 137–138
crystallographic slip, 132
dislocations. *See* Dislocations
Peach–Koehler equation, 151–154
LS coupling, 199–200

M

Madelung constant, 17
Madelung energy, 16
Magnetic quantum number, 5
Magnetite, 65–67, 203–204
Magnetocrystalline anisotropy, 66
Manganite, 75–76
Melting temperatures, 33
Metal-deficient oxides, 94–95, 98–102
Metallic bonds, 31–32
cohesive energy of metals, 31–32
Metallic structures, 39–40
Metals, point defects in. *See* Point defects
Mixed dislocation, 135–137
Monoclinic zirconia, 63
Mott–Hubbard materials, band structures of, 201
Mott–Hubbard theory, 202

N

Neutral atoms, energy for forming the ions from, 13–15
Nonstoichiometric ionic crystals, 96–107
defect structures in, 94–95

O

Octahedral coordination, crystal field in, 47–50, 47*f*, 48*f*
Octahedral field, Jahn–Teller effect in, 55–56
Octahedral site stability energy (OSSE), 52–53
One-degree-of-freedom boundary, 171
Orbital quantum number, 5
Ostwald ripening process, 194
Oxides, defect equilibria in, 113–126
Oxygen sensors, 120–121
Oxygen stoichiometry, 77
Oxygen-deficient oxides, 97–98

P

Pairing energy, 49–50, 56–58
Parsimony, rule of, 46
Partial dislocations, 160–161, 160*f*
 equilibrium separation between, 162–167
 Frank partial dislocations, 165
 Shockley partial dislocation, 161–162, 165
Partially coherent/semicoherent boundary, 175–176, 176*f*
Pauli exclusion principle, 9–10, 30
Pauli principle, 15
Pauling's rules, 41–46
 fifth rule, 46
 first rule, 41–42
 fourth rule, 46
 second rule, 42–44
 third rule, 44–45
Peach–Koehler equation, 151–154
Penetration and shielding of electrons, 7–8
Perovskite ABO_3, structure and the related feature of, 67–80, 67*f*
 $CaCu_3Ti_4O_{12}$, 77–79, 78*f*
 effect of defect, 69
 Jahn–Teller effects, 69
  $BiFeO_3$, 69–70
  Sr_2FeMoO_6, 70–74
 $LaBa_2Cu_3O_7$, 76–77, 76*f*
 manganite, 75–76
 size effects, 67–68
 $SrBaNb_2O_6$, 79–80, 79*f*
 $YBa_2Cu_3O_{7-x}$, 76–77
Perovskite structure, 67–69, 75
Plastic flow, 132
 associated with slip of dislocation motion, 138–142
Point defects
 in binary metal oxides, 92–94
 defect association, 126–127
 defect equilibrium in ionic solids, 109–126
  defect equilibria in oxides, 113–126
  equilibrium concentration of stoichiometric oxides, 109–112
 defect reactions, 95–96
  electroneutrality, 96
  mass balance, 95
  preservation of regular lattice sites ratio, 95
 equilibrium concentration of, 86–92
 formulation of defect equations, 96–109
  extrinsic nonstoichiometry, 107–109
  nonstoichiometric ionic crystals, 96–107
  stoichiometric ionic crystals with internal disorder, 96
 in ionic solids, 92–127
 in metals, 83–92
  factors affecting the solubility of impurity atoms, 85–86
  solid solutions, 83–85
 in nonstoichiometric ionic crystals, 94–95
 in stoichiometric ionic crystals, 94
Polaron, 204
Potential energy, 33–34, 34*f*, 129, 150–151
 attractive, 17
 repulsive, 17
Potential-well concept, effect of bonding on material properties by means of, 32–35
 elastic modulus, 32–33
 melting temperatures, 33
 thermal property, 33–35
Primary bonds, 13
Principal quantum number, 5, 7
Probability function, 5–6, 8
P-type metals, 202–203, 203*f*
Pure solid FCC metal, calculating surface free energy of, 180–181

Q

Quantum number, 5, 9–10

R

Radial distribution function, 5–6, 6*f*, 8–9
Radius ratio rule, 41
Relaxor ferroelectrics, 79–80
Repulsive energy, 18, 28
Repulsive potential energy, 17
Resonance effect, 30
Russell-Saunders coupling, 199

S

Schmid factor, 137–138
Schottky defect, 106–107, 109–111
Schottky disorder, 94, 96
 oxide MO with, 113–126
Schrödinger's wave equation, 3
Screw dislocation, 136
 strain energy of, 146–147
 stress field of, 143–144
Secondary bonds, 13
Self-interstitials, 83
Semicoherent boundary, 175–176

Sessile dislocation, 165
Shielding of electrons, 7–8
Shockley partial dislocations, 161–162, 165
Silicon carbide, 62–63, 62*f*
Sintering, 191–194
  driving force for, 194–197
Slip
  versus climb, 135–136
  crystallographic, 132
  defined, 132
Sodium chloride, 16
  structure of, 16*f*
Solid solutions, 83–85
  interstitial, 85–86
  substitutional, 85
Space lattice, 37–39
Spectroscopic notation, 199–200
Spin, 5
Spinel (AB_2O_4), structure and the related feature of, 64–67, 65*f*
  $LiMn_2O_4$, 65
  magnetite, 65–67
Spin–orbit couplings, 3, 10
Sr_2FeMoO_6, 70–74
  structure of, 70*f*
$SrBaNb_2O_6$, 79–80, 79*f*
Stabilized zirconia, 63–64
Stacking faults, 161–162, 164–165
  in FCC crystals, 161–162
Stoichiometric ionic crystals
  defect structures in, 94
  with internal disorder, 96
Stoichiometric oxides, equilibrium concentration of, 109–112
Strontium barium niobate $Sr_{1-x}Ba_xNb_2O_6$ (SBN), 79
Structures of crystalline crystals, 37
  atoms and ions, arrangements of, 37–39
  ionic and covalent structures, 59–80
    perovskite ABO_3, 67–80, 67*f*
    silicon carbide, 62–63, 62*f*
    spinel (AB_2O_4), 64–67, 65*f*
    TiO_2, 63, 63*f*
    zinc oxide, 61–62, 61*f*
    ZrO_2, 63–64, 64*f*
  ionic structures, 40–46
    Pauling's rules. *See* Pauling's rules
  ionic structures, structural distortion in, 46–59
    arising from asymmetric electron density/inert pair effect, 59
    crystal field stabilization energy (CFSE), 51–53
    effects of the d electron configuration, 53–55
    Jahn–Teller distortions, 55–59
    octahedral coordination, crystal field in, 47–50, 47*f*, 48*f*
    tetrahedral coordination, crystal field in, 50–51, 51*f*
  metallic structures, 39–40
Substitutional solid solution, 83
Surface free energy, 176–179
Surface stress, 176–179
Surface tension, 163, 176–179
Surface tension force balance, 186*f*, 188*f*, 195*f*

T

t_2 orbitals, 50, 58, 58*f*
Taylor series, 32
Tetrahedral coordination, crystal field in, 50–51, 51*f*
Tetrahedral field, Jahn–Teller effect in, 58–59
Thermal expansion, 34–35, 62–63
Thermal property, 33–35
Thermodynamic viewpoint of bond energy, 15–29
  Born–Haber cycle, 29
  electrostatic energy, 15–18
  forces between ions in an ionic crystal, 18–22
  lattice energy, approximation of, 18–22
  repulsive energy, 18
  van der Waals attractive energy, 22–26
    field produced by a dipole, 22–24
    induced dipoles, 24–26
  zero-point energy, 26–28
3-d transition metal, 201–202
Tilt boundary, 170–171
TiO_2, structure and the related feature of, 63, 63*f*
Transition metal, 9–10
Trimerons, 66–67
Tungsten bronzes (TBs), 79
Twist boundary, 173
Two-degree-of-freedom boundary, 169
Two-dimensional and three-dimensional imperfections in solids, 169
  equilibrium vacancy concentration, 194–197
  free surface, 179–181

calculating surface free energy of pure solid FCC metal, 180–181
grain boundaries, 169–173
coincidence site lattice, 173
tilt boundary, 171–173
twist boundary, 173
interface curvature, effect of
on equilibrium pressure between two phases, 188–191
on equilibrium solubility between two phases, 191–194
interfaces of phases, 181–188
equilibrium shapes of grains, 182–183
presence of second phases, 184–188
wetting, 181–182
interphase boundaries in solids, 174–176
coherent boundary, 174–175
incoherent boundary, 175
partially coherent/semicoherent boundary, 175–176
surface free energy, 176–179
surface stress, 176–179
surface tension, 176–179

U

Unit cell, 38–39, 39*f*

V

van der Waals attractive energy, 22–26
field produced by a dipole, 22–24
induced dipoles, 24–26
Verwey transition, 66, 66*f*
Vibrational energy, 26–27
Virtual force, 148–149, 148*f*

W

Wetting, 181–182

Y

$YBa_2Cu_3O_{7-x}$, 76–77

Z

Zero-point energy, 26–28, 33–34
Zigzag APBs, 72–73
Zinc oxide, structure and the related feature of, 61–62, 61*f*
ZrO_2, structure and the related feature of, 63–64, 64*f*

Kolibri 360 GmbH
Ferdinand-Jühlke-Straße 7
99095 Erfurt, Deutschland
produktsicherheit@kolibri360.de